LONDON : GEOFFREY CUMBERLEGE
OXFORD UNIVERSITY PRESS

JOURNAL OF A VISIT
TO LONDON AND THE CONTINENT
BY HERMAN MELVILLE
1849—1850

EDITED BY

Eleanor Melville Metcalf

HARVARD UNIVERSITY PRESS
CAMBRIDGE, MASSACHUSETTS
1948

COPYRIGHT, 1948
BY THE PRESIDENT AND FELLOWS OF
HARVARD COLLEGE

PRINTED IN THE UNITED STATES OF AMERICA

PREFACE

The complete text of Melville's *Journal of 1849* appears here for the first time. The Journal was a private record that was never intended for publication, but what its author never purposed, its humanity claims for it now. In order to make an easily readable book, but at the same time not to smooth away all eccentricities of the original, I have tried to pursue a middle course and have left Melville's characteristic usages where I felt they would be no obstacle to the eye.

To read correctly what Mrs. Hawthorne said "requires second sight to decipher, the handwriting being apparently 'writ in water,'" is a serious problem only where the exactness of the word is important to the meaning. In such cases further knowledge of the conditions involved or some curious unexpected fact suddenly shining out of the thickets of antiquity has often solved the problem. But in the few places where there still remain undecipherable words I have discussed them in the notes.

The evident haste with which Melville often wrote, his ignorance of foreign languages, and his curiously defective ear (curious in one who was master of such sonorous prose) account for many vagaries of spelling. These vagaries — original without uniformity — have been corrected and modernized wherever they would have presented a hardship to the reader. But since complete uniformity would have taken a good deal of characteristic flavor out of the text, I have left in as much of the original as I could. Capitals have been supplied where he omitted them; but nouns which he capitalized have been left as he wrote them. Certain fashions of the period, such as M^r for Mr., have not been followed; but ampersands have been kept wherever they occur, since they are not so numerous as to distract the modern eye.

Melville's punctuation has been modernized where the sentence structure would otherwise obscure the meaning. Jottings without sentence form are left separated by dashes, as in the original. But I have omitted some dashes when they were obviously superfluous and hindered rather than helped the meaning. His lack of uniformity in punctuation might tempt one to think that Melville shared De Quincey's contempt for "crotchety authors supersensitively fastidious in matters of punctuation," were it not for a remark to his publishers, Dix and Edwards, accompanying his proofs of *Piazza Tales* in 1856, "There seem to have been a surprising profusion of commas in the proofs. I have struck them out pretty much, but hope that someone who understands punctuation better than I do, will give the final hand to it."

His deletions, his changes of wording (such as *close-reefed* for *double-reefed*), and variant readings of questionable words are given in the notes following the text. His wife's penciled notes are reproduced in footnotes. As the reader will see, Melville was eccentric in his dating. Usually I have restored the correct date and accounted for his vagaries in the notes. The dates dividing entries have been harmonized and given a uniform position on the page. In explanation of his cancellations, it is evident that Melville thought better of some unflattering remarks he made about certain people and occasions (in one case, really venomous remarks), for he heavily inked them out. But the modern technique of exploratory photography has brought to light almost all the banished words. There seems to be no reason now for excluding them. I am indebted to Miss Belle Da Costa Greene, Director of the Pierpont Morgan Library, and to the library's photographer, Mr. Brewer, for the results.

Students who wish to see the uncorrected text may consult the two original notebooks and my typed version in the Harvard College Library. The first notebook is a thin, fragile paper book containing only the account of the voyage out, "Journal of a Voyage from New York to London 1849." The second is a firmly bound red book, which he must have bought on his arrival in London. "I commence this Journal at 25 Craven Street, Strand, at 6½ P.M. on Wednesday Nov. 7th 1849 — being just arrived from a chop House, and feeling like it." He begins it with an account of the last few hours aboard ship, the landing, and incidents of the trip to London.

The student will find in these a list of addresses belonging to the Pittsfield period, 1850–1863, and the beginning of his journal of 1856–1857, the *Journal Up The Straits,* for obvious reasons omitted from this edition.

All references to Melville's works are to the Constable Edition, 1922–1924, with the exception of the *Journal Up The Straits,* published in 1935 by The Colophon, New York.

I wish here to acknowledge my debt to Mr. William Henry McCarthy formerly of the Houghton Library, Harvard, whose indefatigable zeal has lightened my labors in countless practical ways; and to Mr. Jay Leyda, who has shared with extraordinary generosity the results of his recent painstaking research in preparation for his own book on Melville.

If I were to name all who have given me of their time and the results of their learning, labors, and insights, I should add a long list of Melville students, specialists in particular fields, and members of the staffs of the Harvard College Library, the New York Public Library, the Albany Public Library, the Massachusetts Historical Society, the Pierpont Morgan Library, the Library of Congress, the Old Dartmouth Historical Society and Whaling Museum, New Bedford; the Dukes County Historical Society, Edgartown; the Fogg Art Museum, Cambridge; the

Berkshire Athenaeum, the Boston Athenaeum, the British Museum, and the Central Library of Portsmouth, England.

I here thank them all. Each knows (though perhaps cannot fully realize) what each has done for this edition of the *Journal*. To the growing body of Melville scholars, and to that part of the larger public that is curious about this no-longer-neglected American writer of the nineteenth century, this annotated edition of Melville's *Journal of 1849* is offered by his granddaughter,

ELEANOR MELVILLE METCALF

*

[xi]

CONTENTS

ILLUSTRATIONS

[xv]

Con. Here's to you now,
And you shall find his travel has not stopt him
As you suppose, nor alter'd any freedome,
But made him far more clear and excellent;
It draws the grosness off the understanding,
And renders active and industrious spirits:
He that knows most mens manners, must of necessity
Best know his own, and mend those by example:
'Tis a dull thing to travell like a Mill-horse,
Still in the place he was born in, lam'd and blinded;
Living at home is like it: pure and strong spirits
That like the fire still covet to fly upward,
And to give fire as well as take it; cas'd up, and mewd here
I mean at home, like lusty metled Horses,
Only ty'd up in Stables, to please their Masters,
Beat out their fiery lives in their own Litters,
Why do not you travel Sir? *

Beaumont and Fletcher,
The Queen of Corinth, act II, scene I

* Passage marked by Melville in his copy of the *Plays,* bought in London, inscribed with his name, December, 1849, and read on the trip home (New Year's Day, at sea).

[xvi]

INTRODUCTION

For seventeen weeks and two days of 1849-50, Herman Melville, then a young man of thirty with an already established reputation in the literate English-speaking world, kept a journal of a trip to Europe.

The immediate purpose of his trip was to find a publisher for *White Jacket*, perhaps necessary since the comparative unpopularity of *Mardi,* and the death of his brother, Gansevoort, who had marketed *Typee* for him in London. But that he had other motives less obviously practical, he freely revealed to friends before he sailed and discussed with some of his fellow travellers aboard the ship.

This journal was written at a time when he was beginning to feel the stirrings of depths of which he could not then know the outcome. It exhibits poignantly the conflict between his eagerness to experience for himself the wonders of the magical East and his homesick longing for his young wife and son in America.

These two main streams of feeling underlie the record of a quick and varied response to historical and natural scenes, rich accumulations of the arts, social life, and literary and philosophical discussions. That his experiences never remained those of the typical tourist, is shown by the use he made of them afterwards and in his designation of himself as "a pondering man" in a lonely moment of his travels.

Forced by lack of funds to forego any further thought of a "glorious eastern jaunt" at this time, he returned to America and set to work to earn his family's living in the only way he knew, the production of another book. For the present, the East must suggest its magic to him in such books as *Anastasius* and *Vathek*, which he acquired on this trip.

But to go back to the autumn of 1849, among the City Items of the *New York Daily Tribune,* Thursday, October 11th, appeared the notice that "Herman Melville has sailed for London in the packet-ship Southampton. It is his intention to spend a twelvemonth abroad."

The *Home Journal* also gave notice of his departure. November 8, 1849, under "Brevities," it carried the following item:

> Herman Melville sailed a few days since for England, on a first visit to that land of things perfected. With his genius, the popularity of his books in England, and the extraordinary charm of his narrative powers in conversation, we predict for him an 'open sesame' through the most difficult portals of English society.

He had made the preparations of one who was far from considering himself negligible: he had asked for and received from Evert and George Duyckinck, from his father-in-law Judge Shaw, and through him, from Edward Everett, letters of introduction to notables in London. And in addition, he had been given other letters by interested friends, whether asked or unasked is not known. He had followed the usual custom of travelers in listing possible acquaintances, and making notes of sights, lodgings, and eating places.

In a book given him by his friend, Dr. Augustus Gardner, *Old Wine In New Bottles or Spare Hours of a Student in Paris,* he wrote in the back cover,

Museum Dupuytren 69
'charmante'
Pere la Chaise
The Chaumiere (dance house) 309

This bears on the title page a quotation from Johnson: "The use of travelling is to regulate imagination by reality; and instead of thinking how things may be, to see them as they are."

Melville might have written, "The use of travelling is to regulate reality by imagination; and instead of *thinking,* to *see* things as they are."

For "to see things as they are," finally necessitated for him their transmutation by the imagination. In this journal the stuff of a few weeks' experience appears, or is suggested, or may be followed through the books he read, before the imagination had intensified its elements and given them ordered form. In the notes which follow the text may be seen some of the connections between this private record and many of his later works, even poems published in the last year of his life. His narrative powers in the purely descriptive portions make a lively enjoyment of time, place, and person, and reflect with unconscious clarity the personality of the writer.

THE VOYAGE

HERMAN MELVILLE'S
JOURNAL
OF A VOYAGE FROM
NEW YORK TO LONDON

1849

After a detention of three or four days, owing to wind & weather, with the rest of the passengers I went on board the tugboat Goliath about 12 ½P.M. during a cold violent storm from the West. The "Southampton" (a regular London liner) lay in the North River. We transferred ourselves aboard with some confusion, hove up our anchor, & were off. Our pilot, a large, beefy-looking fellow, resembled an oysterman more than a sailor. We got outside the "Narrows" about 2 o'clock; shortly after, the "tug" left us & the Pilot. At half past 5 P.M. saw the last of the land, with our yards square, & in half a gale.

As the ship dashed on, under double-reefed top-sails, I walked the deck, thinking of what they might be doing at home, & of the last familiar faces I saw on the wharf — Allan was there, & George Duyckinck, and a Mr. McCurdy, a rich merchant of New York, who had seemed somewhat interested in the prospect of his son (a sickly youth of twenty bound for the grand tour) being my roommate. But to my great delight, the promise that the Captain had given me at an early day, he now made good; & I find myself in

[3]

the undivided occupancy of a large stateroom. It is as big almost as my own room at home; it has a spacious berth, a large washstand, a sofa, glass &c. &c. I am the only person on board who is thus honored with a room to himself. I have plenty of light, & a little thick glass window in the side, which in fine weather I may open to the air. I have looked out upon the sea from it, often, though not yet 24 hours on board.

Friday, October 12

Walked the deck last night till about eight o'clock; then made up a whist party & played — till one of the number had to visit his room from sickness. Retired early & had a sound sleep. Was up betimes, & aloft, to recall the old emotions of being at the masthead. Found that the ocean looked the same as ever. Have tried to read, but found it hard work. However, there are some very pleasant passengers on board with whom to converse. Chief among these is a Mr. Adler, a German scholar, to whom Duyckinck introduced me. He is author of a formidable lexicon, (German & English); in compiling which he almost ruined his health. He was almost crazy, he tells me, for a time. He is full of the German metaphysics, & discourses of Kant, Swedenborg, &c. He has been my principal companion thus far. There is also a Mr. Taylor among the passengers, cousin to James Bayard Taylor, the pedestrian traveller. He is full of fun — or rather *was* full of it. Just at this moment I hear

[4]

Herman Melville

Journal
Of a Voyage from
New York to London
1849

Thursday   After a detention of three or
Oct 11th four days, owing to wind & weather,
with the rest of the passengers ~~~~
I went on board the tug-boat
Goliath about 12½ P.M. during
a cold violent storm from the
West. The "Southampton" (a regular
London liner) lay in the North river.
We transferred ourselves aboard with
some confusion, bore up our anchor,
& were off. Our pilot, a large, beefy
looking fellow resembled an oysterman more than a sailor. We got outside
the "narrows" about 2 Oclock; shortly after,
the "tug" left us & the Pilot. At half
past 5. P.M., saw the last of the land,
with our yard square, & in half a gale.

Facsimile — see page 3

his mysterious noises from the stateroom next to mine. Poor fellow! he is seasick. As yet there have been but few thus troubled, owing to pleasant weather. There is a Scotch artist on board, a painter, with a most unpoetical-looking only child, a young one all cheeks & forehead, the former preponderating. Young McCurdy I find to be a lisping youth of genteel capacity, but quite disposed to be sociable. We have several Frenchmen & Englishmen. One of the latter has been hunting, & carries over with him two glorious pairs of antlers (moose) as trophies of his prowess in the woods of Maine. We have also, a middle-aged English woman, who sturdily walks the deck, & prides herself upon her sea legs, & being an old tar.

Saturday, October 13

Last evening was very pleasant. Walked the deck with the German, Mr. Adler, till a late hour, talking of "Fixed Fate, Free will, foreknowledge absolute" &c. His philosophy is Coleridgean: he accepts the Scriptures as divine, & yet leaves himself free to inquire into Nature. He does not take it that the Bible is absolutely infallible, & that anything opposed to it in Science must be wrong. He believes that there are things *out* of God and independent of him, — things that would have existed were there no God: — such as that two & two make four; for it is not that God so decrees mathematically, but that in the very nature of things, the fact is thus.

[5]

Rose early this morning, opened my bull's-eye window, & looked out to the East. The sun was just rising, the horizon was red; a familiar sight to me, reminding me of old times. Before breakfast went up to the masthead, by way of gymnastics. About 10 o'clock A.M. the wind rose, the rain fell, & the deck looked dismally enough. By dinner time, it blew half a gale, & the passengers mostly retired to their rooms, seasick. After dinner, the rain ceased, but it still blew stiffly, & we were slowly forging along under close-reefed topsails — mainsail furled. I was walking the deck, when I perceived one of the steerage passengers looking over the side; I looked too, & saw a man in the water, his head completely lifted above the waves, — about twelve feet from the ship, right abreast the gangway. For an instant, I thought I was dreaming; for no one else seemed to see what I did. Next moment, I shouted "Man overboard!" & turned to go aft. The Captain ran forward, greatly confused. I dropped overboard the tackle fall of the quarter-boat, & swung it towards the man, who was now drifting close to the ship. He did not get hold of it, & I got over the side, within a foot or two of the sea, & again swung the rope towards him. He now got hold of it. By this time, a crowd of people — sailors & others — were clustering about the bulwarks; but none seemed very anxious to save him. They warned *me*, however, not to fall overboard. After holding on to the rope about a quarter of a minute, the man let go of it, & drifted astern under the mizzen chains. Four or five of the seamen jumped over into the chains & swung him more ropes. But his conduct was unaccountable; he could have saved himself, had he been so minded. I was struck by the expression of his face in the water. It was merry. At last he drifted off under the ship's counter, & all hands cried "He's gone!" Running to the taffrail, we saw him again, floating off — saw a few bubbles & never saw him again. No boat was lowered, no sail was shortened, hardly any noise was made. The man drowned like a bullock. It afterwards turned out, that he was crazy, & had jumped overboard. He had declared he would do so several times; & just before he *did* jump, he had tried to get possession of his child, in order to jump into the sea with the child in his arms. His wife was miserably sick in her berth. The Captain said that this was the fourth or fifth instance he had known of people jumping overboard. He told a story of a man who did so, with his wife on deck at the time. As they were trying to save him, the wife said it was no use; & when he was drowned, she said "there were plenty more men to be had." Amiable creature! By night, it blew a terrific gale, & we hove to.

Miserable time! nearly every one sick, & the ship rolling & pitching in an amazing manner. About midnight, I rose & went on deck. It was blowing horribly — pitch dark, & raining. The Captain was in the cuddy, & directed my attention "to those fellows" as he called them, — meaning several "Corposant balls" on the yardarms & mastheads. They were the first I had ever seen, & resembled large, dim stars in the sky.

A regular blue devil day. A gale of wind, & every one sick. Saloons deserted, & all sorts of nausea noise heard from the staterooms. Taylor, McCurdy, & Adler all in their berths — & I alone am left to tell the tale of their misery. Read a little in Mrs. Kirkland's European tour. Like it. She is a spirited, sensible, fine woman. Managed to get through the day somehow, by reading & walking the deck, though the last was almost as much as my neck was worth. I forgot to say that shortly after the loss of the crazy man (a Dutchman by the way) some of the steerage passengers came aft & told the Captain that there was another crazy man, an Englishman in the steerage. This morning, coming on deck, I saw a man leaning against the bulwarks, whom I immediately took for a steerage passenger. He stopped me, & told me to look off & *see the steamers.* So I looked for about five minutes, — straining my eyes very hard, but saw nothing. — I asked the 2d Mate whether *he* could see the steamers; when he told me that my informant was the crazy Englishman. All the morning this poor fellow was on deck, crying out at steamers, boats, &c &c. I thought that his mad feelings found something congenial in the riot of the raging sea. In the evening, he forced his way into the dining saloon, & struck the Steward, who knocked him down, & dragged him forward. We have made no progress for the last 36 hours; wind ahead, from the Eastward. The crazy man turns out to be afflicted with delirium tremens, consequent upon keeping drunk for the last two months. He is very earnest in his enquiries after a certain *Dr. Dobbs*. Saw a lady with a copy of *Omoo* in her hand two days ago. Now & then she would look up at me, as if comparing notes. She turns out to be the wife of a young Scotchman, an artist, going out to Scotland to sketch scenes for his patrons in Albany, including Dr. Armsby. He introduced himself to me by mentioning the name of Mr. Twitchell who painted my portrait gratis. He is a very unpretending young man, & looks more like a tailor than an artist. But appearances are &c. —

Monday, October 15

The gale has gone down, & we have fine weather. By noon the passengers were pretty nearly all on deck, convalescent. They seem to regard me as a hero, proof against wind & weather. My occasional feats in the rigging are regarded as a species of tight-rope dancing. Poor Adler, however, is hardly himself again. He is an exceedingly amiable man, & a fine scholar whose society is improving in a high degree. This afternoon Dr. Taylor & I sketched a plan for going down the Danube from Vienna to Constantinople; thence to Athens on the steamer; to Beyrouth

& Jerusalem — Alexandria & the Pyramids. From what I learn, I have no doubt this can be done at a comparatively trifling expense. Taylor has had a good deal of experience in cheap European travel, & from his knowledge of German is well fitted for a travelling companion through Austria & Turkey. I am full (just now) of this glorious *Eastern* jaunt. Think of it! Jerusalem & the Pyramids — Constantinople, the Aegean, & old Athens! The wind is not fair yet, & there is much growling consequently. Drank a small bottle of London Stout today for dinner, & think it did me good. I wonder how much they charge for it? I must find out; & not go through the sad experience that "Powell" did (as he says).

Tuesday, October 16

Beautiful weather, but wind against us. Passengers all better, & quite lively; excepting young McCurdy with a touch of the ague, and a lady, who seems quite ill. Read little or nothing, but lounged about. The sea has produced a temporary effect upon me, which makes me for the time incapable of anything but vegetating. What's little Barney * about? Where's Orianna? †

Wednesday, October 17

Fine weather, quite warm & sunny. The decks lively, the ladies lively, the Captain lively, & the ship now going her course. Spent a good part of the

---

* *Macky.*   † *Lizzie.*

Herman Melville

day aloft with Adler, in conversation. In the evening had a sort of concert. An Irish lady, an opera singer they say, leading off with a guitar & her voice.

Delightful day, & the ship getting on famously. Spent the entire morning in the main-top with Adler & Dr. Taylor, discussing our plans for the grand circuit of Europe & the East. Taylor, however, has communicated to me a circumstance, that may prevent him from accompanying us — something of a pecuniary nature. He reckons our expenses at $400.

No events; spent the morning in lounging & reading; and after a hand at cards, retired.

Newfoundland weather — foggy, rainy &c. Read account of Venice in Murray. Cleared up in the afternoon — passengers played shuffleboard on the quarter-deck. For the first time promenaded with some of the ladies — a Mrs —— of Monmouthshire, England, & a Miss Wilbur (I think) of New York. The former is flat: the latter is of a marriageable age, keeps a diary & talks about "winning souls to Christ." In the evening for the first time went into the Ladies' Saloon, & heard Mrs. Gould, the opera lady, sing. There was quite a party — the saloon is

[11]

gilt & brilliant, & as the ship was going on quietly, it seemed as if I were ashore in a little parlor or cabinet. Where's Orianna? * How's little Barney? † Read a chapter in Pickwick & retired pretty early. Towards morning was annoyed by a crying baby adjoining.

Sunday, October 21

Rainy — near the Banks. Can not remember what happened today. It came to an end somehow.

Monday, October 22

Clear & cold; wind not favorable. I forgot to mention, that *last night* about 9½ P.M. Adler & Taylor came into my room, & it was proposed to have whisky punches, which we *did* have, accordingly. Adler drank about three tablespoons full — Taylor 4 or five tumblers &c. We had an extraordinary time & did not break up till after two in the morning. We talked metaphysics continually, & Hegel, Schlegel, Kant &c were discussed under the influence of the whiskey. I shall not forget Adler's look when he quoted La Place the French astronomer — "It is not necessary, gentlemen, to account for these worlds by the hypothesis" &c. After Adler retired Taylor & I went out on the bowsprit — splendid spectacle. It came on calm in the evening, & we await a favorable shift of wind.

* *Lizzie.*   † *Macky.*

[12]

Tuesday, October 23

On gaining the deck this morning, was delighted to find a fair wind. It soon blew stiff, & we scudded before it under double-reefed topsails, & mainsail hauled up. Running about 14 knots all day. Every one in high spirits. Captain told a rum story about a *short skipper* and a *long mate* in a little brig, & throwing overboard the barrels of beef & turpentine &c.

Wednesday, October 24

Fair wind still holds on; at 12 M. supposed to be half way over. Saw several land birds — very tame, lighted on deck — caught one.

Thursday, October 25

A fair wind — good deal of rain. About noon saw a ship on the other tack. She showed her colors & proved a Yankee. The first vessel that we had seen so near. She excited much interest. By evening blew a very stiff breeze, & we dashed on in magnificent style. Fine moonlight night, & we rushed on through snow-banks of foam. McCurdy invited Adler, the Doctor & I into his room & ordered champagne. Went on deck again, & remained till near midnight. The scene was indescribable. I never saw such sailing before.

Friday, October 26

Fair wind still. Towards noon came on calm, with a gawky sea. The ship rolled violently, & many comical scenes ensued among the passengers. Breezed up again in the afternoon, & we went on finely. For a few days past, Adler & I have had some "sober second thoughts" about our grand Oriental & Spanish tour with Taylor. But tonight, the sight of "Bradshaw's Railway & Steamer Guide" showing the marvellous ease with which the most distant voyages may now be accomplished has revived — at least in *my* mind, — all my original enthusiasm. Talked the whole thing over again with Taylor. Shall not be able to decide till we get to London.

Saturday, October 27

Steered our course on a wind. I played Shuffleboard for the first time. Ran about aloft a good deal. McCurdy invited Adler, Taylor & I to partake of some *mulled wine* with him, which we did, in my room. Got — all of us — riding on the German horse again — Taylor has not been in Germany in vain. After another curious discussion between the Swede & the Frenchman about Lamartine & Corinne, we sat down to whist, & separated at about 3 in the morning.

Sunday, October 28

Came on a strong breeze & lasted all day. Ship going about 12 miles an hour — begin to talk of port. Decks very wet, & hard work to take exercise ("Where dat old man"?)* Read a little, dozed a little, & to bed early.

* Macky's baby words.

Wet & foggy, but a fair fresh breeze — 12 knots an hour. Some of the passengers sick again. In the afternoon tried to create some amusement by arraigning Adler before the Captain on a criminal charge. In the evening put the Captain in the Chair, & argued the question "which was best, a Monarchy or a republic?" — Had some good sport during the debate — the Englishmen wouldn't take part in it though. After chat & Stout with Monsieur Moran & Taylor, went on deck, & found it a moonlight midnight. Wind astern. Retired at 1 A.M.

Tuesday, October 30

Glorious day — Capital cakes for breakfast. ("Where dat old man"?)* Saw a land bird. Weather beautifully clear. For the first time in five days got our observation. Find ourselves heading right into the middle of the Channel — the Scilly Islands out of sight to the North. Played Shuffleboard with Taylor & the ladies. Had a superb dinner, which we all relished amazingly. Drawing near port with a fine fair wind makes passengers feel generous. A good deal of wine and porter on table. A magnificent night — but turned in very early.

Wednesday, October 31

Fair, fresh wind still holds. Coming on deck in the morning saw a brig close to — & two or three ships.

* Macky's baby talk.

If the wind holds we shall make the Lizzard Light this evening, probably. May be in Portsmouth tomorrow night. All hands in high spirits. Had some mulled Sherry in the evening from McCurdy. Up late, expecting it to be the last night.

Thursday, November 1

Just three weeks from home, and made the land — Start Point — about 3 P.M. — well up Channel — past the Lizzard. Very fine day — great number of ships in sight. Through these waters Blake's & Nelson's ships once sailed. Taylor suggested that he & I should return McCurdy's civilities. We did, and Captain Griswold joined and ordered a pitcher of his own. The Captain is a very intelligent & gentlemanly man — converses well & understands himself. I never was more deceived in a person than I was in him. Retired about midnight. Taylor played a rare joke upon McCurdy this evening, passing himself off as Miss Wilbur, having borrowed her cloak &c. They walked together. Shall see Portsmouth tomorrow morning.

Friday, November 2

Wind from the East — ahead. Clear & beautiful day but every one grievously disappointed. I think I shall get off at Portsmouth, instead of going round. May be in tonight, after all. Spoke a Portsmouth Pilot boat, but took no pilot. Made the Bill of Portland — from which the Portland stone is got. Melancholy looking voyage, white cliffs indeed! In the evening played chess, & talked metaphysics [with] my learned friend till midnight.

Woke about 6 o'clock with an insane idea that we were going before the wind, & would be in Portsmouth in an hour's time. Soon found out my mistake. About eight o'clock took a pilot, who brought some papers two weeks old. Made the Isle of Wight about 10 A.M. High land — The Needles. Wind ahead & tacking. Get in tonight or tomorrow — or next week or year. Devilish dull, & too bad altogether. Continued tacking all day with a light wind from West. Isle of Wight in sight all day & numerous ships. One of our steerage passengers left in the Pilot Boat. Rum scene alongside with the boat. In the evening all hands in high spirits — Played chess in the ladies' saloon — another party at cards; good deal of singing in the gentlemen's cabin & drinking — very hilarious & noisy — Last night, every one thought. Determined to go ashore at Portsmouth. Therefore prepared for it — arranged my trunk to be left behind — put up a shirt or two in Adler's carpet bag & retired pretty early.

Sunday, November 4

Looked out of my window first thing upon rising & saw the Isle of Wight again — very near — ploughed fields &c. Light head wind — expected to be in a little after breakfast time. About 10 A.M. rounded the Eastern end of the Isle, when it fell flat calm. The town in sight by telescope. Were becalmed about three or four hours. Foggy, drizzly; long faces at dinner — no porter bottles. Wind came from the West at last. Squared the yards & struck away for Dover — distant 60 miles. At 6 o'clock (evening) passed Dungeness — then saw the Beachy Head light. Close reefed the topsails so as not to run too fast. Expect now to go ashore tomorrow morning early at Dover — & get to London via Canterbury Cathedral. Mysterious hint dropped me about my green coat. Talked with the Pilot about the perils of the Channel. He told a story of running down a brig in a steamer &c. It is now eight o'clock in the evening. I am alone in my stateroom — lamp in tumbler. Spite of past disappointments I *feel* that this is my last night aboard the Southampton. This time tomorrow I shall be on land, & press English earth after the lapse of ten years — *then* a sailor, *now* H. M. author of "Peedee" "Hullabaloo" & "Pog-Dog." For the last time I lay aside my "*log*," to add a line or two to Lizzie's letter — the last I shall write aboard. ("Where dat old man?" — "Where books?")*

Oct. 15. Small bottle of stout (Dinner)
One bottle   ”
two bottles (one at night)
one bottle (one at night)
  „   „   afternoon
  „   „   afternoon

* First words of baby Malcolm's.

[18]

LONDON
AND
THE CONTINENT

I commence this Journal at 25 Craven Street, Strand, at 6½ P.M. on Wednesday November 7th, 1849 — being just arrived from dinner at a chop House, and feeling like it.*

Monday, November 5

Having at the invitation of McCurdy cracked some Champagne with him, I retired about midnight to my stateroom; & at 5 in the morning was wakened by the Captain in person, saying we were off Dover. Dressed in a hurry, ran on deck, & saw the lights ashore. A cutter was alongside, and after some confusion in the dark, we got off in her for the shore. A comical scene ensued, the boatman saying we could not land at Dover, but only at Deal. So to Deal we went, & were beached there just at break of day. Some centuries ago a person called Julius Caesar jumped ashore about in this place, & took possession. It was Guy Fawke's Day also. Having left our baggage (that is Taylor, Adler & self) to go round by ship to London, we were wholly unencumbered; & I proposed walking to Canterbury — distant 18 miles, for an appetite to breakfast. So, we strode through this quaint old town of Deal — one of the Cinque Ports, I believe, and soon were in the open country.

* For the Journal of the voyage out, see the small paper book.

[21]

A fine autumnal morning & the change from ship to shore was delightful. Reached Sandwich (6 miles) and breakfasted at a tumbledown old inn. Finished with ale & pipes. Visited "Richboro' Castle" — so-called — a Roman fortification near the seashore. An imposing ruin; the interior was planted with cabbages. The walls some 10 feet thick, grown over with ivy. Walked to where they were digging — and saw defined by a trench, the exterior wall of a circus. Met the proprietor — an antiquary — who regaled us with the history of the place. Strolled about the town, on our return, and found it full of interest as a fine specimen of the old Elizabethan architecture. Kent abounds in such towns. At one o'clock took the 2d class (no 3d) cars for Canterbury. The Cathedral is on many accounts the most remarkable in England. Henry II, his wife, & the Black Prince are here — & Becket. Ugly place where they killed him. Fine cloisters. There is a fine thought expressed in one of the inscriptions on a tomb in the nave.

Visited the Dane John in the afternoon, rather evening. Beautiful view of the city & its numerous old churches. The old wall forms a fine promenade. Dined at the "Falstaff" Inn near the Westgate. Went to the theater in the evening, & was greatly amused at the performance: more people on the stage than in the boxes. Ineffably funny, the whole affair. All three of us slept in one room at the inn — odd hole.

Swallowed a glass of ale, & away for the R. R. Station, & off for London, distant some 80 miles. Took the third class cars — exposed to the air — devilish cold riding against the wind. Fine day — people sociable. Passed through Penshurst (P. S.'s place) & Tunbridge (fine old ruin there). Arrived at London Bridge at noon — crossed at once over into the city, & dinner at a chop house in the Poultry — having eaten nothing since the previous afternoon dinner. Went on past St. Paul's to the Strand to find our house. They referred us elsewhere, very full. Secured rooms at last (one for each) at a guinea & a half per week. Very cheap. Went down to the Queen's Hotel to inquire after our ship friends — (on the way green coat attracted attention). Not in. Went to Drury Lane at Julien's Promenade Concerts. (Admittance 1s.) A great crowd, & fine music. In the reading room happened to see "Bentley's Miscellany" with something about Redburn. (By the way, stopped at a store in the Row, & inquired for the book, to see whether it had been published. They offered it to me at a guinea). At Julien's also saw Blackwood's long story about a short book. It's very comical — seemed so, at least, as I had to hurry over it — in treating the thing as real. But the wonder is that the old Tory should waste so many pages upon a thing, which I, the author, know to be trash, & wrote it to buy some tobacco with — On the way home, stopped at

the American Bowling Saloon in the Strand. A good
wash & turned in early.

Wednesday, November 7

Rose at eight, & away we went down into the city &
breakfasted at a "hole in the wall." Then to the
Blackwall R. R. Station for the East India Docks,
after our trunks. After infinite trouble with the
cursed Customs, we managed to get them through.
(Two disconsolates on board the ship.) At five P.M.
arrived home, & dined, & went to see Madame Vestris
& Charles Mathews at the Royal Lyceum Theatre,
Strand. Went into the Gallery (one shilling). Quite
decent people there — fellow going round with a
coffeepot & mugs, crying "Porter, gents, porter!"

Thursday, November 8

Dressed, after breakfast at a Coffee house, & went to
Mr. Bentley's. He was out of town, at Brighton. The
notices of "Redburn" were shown me — laughable.
Stayed awhile, & then to Mr. Murray's. Out of town.
Strolled about, & went into the National Gallery.
Dined with the Doctor & Adler, & after a dark ramble
through Chancery Lane & Lincoln's Inn Fields, we
turned into Holborn, & so to the Princess's Theatre
in Oxford Street. Went into the pit at the half price
— one shilling. The part of a Frenchman was very
well played. So also, skaters on the ice.

Breakfasted late, & went down to Queen's Hotel — saw McCurdy there & Mulligan. Parted from the Doctor & Adler near the Post Office, & went into Cheapside to see the "Lord Mayor's Show" it being the day of the great civic feast & festivities. A most bloated pomp, to be sure. Went down to the bridges to see the people crowding there. Crossed by Westminster, through the Parks to the Edgeware Road, & found the walk delightful — the sun coming out a little, & the air not cold. While on one of the Bridges, the thought struck me again that a fine thing might be written about a Blue Monday in November London — a city of Dis (Dante's) — clouds of smoke — the damned &c. — coal barges — coaly waters, cast-iron Duke &c. — its marks are left upon you, &c.&c.&c.

Stopped in at the Gallery of the Adelphi Theatre, Strand — horribly hot & crowded — good piece though — was in bed by ten o'clock.

At breakfast received a note from Mr. Bentley in reply to mine, saying he would come up from Brighton at any time convenient to me. Wrote him, "Monday noon, in New Burlington St." After breakfast at a Coffee room, Adler went off to Hampton Court & the Dr. to the Botanic Gardens, Regent's Park. For me, I lounged away the day — sauntering through the Temple courts & gardens, Lincoln's Inn, The New Hall, Gray's Inn, down Holborn Hill through

Cock Lane (Dr. Johnson's Ghost) to Smithfield (West). And so on to the Charter House, where I had a sociable chat with an old pensioner who guided me through some fine old cloisters, kitchens, chapels. Lord Ellenborough lies buried hard by the founder. They bury all their dead on their own premises. Duke of Norfolk was confined here in Elizabeth's time for treason. From the Charter House through the Goswell Street Road to Barbican towards London Wall. Asked an officer of the Fire Department where lay St. Swithin's — He was very civil & polite & offered to show me the way in person. "Perhaps you would like to see the way to the house where Whittington was born? Many Londoners never saw it." "Lead on," said I — & on we went — through squalid lanes, alleys, & closes, till we got to a dirty blind lane; & there it was with a slab inserted in the wall. Thence, through the influence of the Fire Officer, I pushed my way through cellars & anti-lanes into the rear of Guildhall, with a crowd of beggars who were going to receive the broken meats & pies from yesterday's grand banquet (Lord Mayor's Day). — Within the hall, the scene was comical. Under the flaming banners & devices, were old broken tables set out with heaps of fowls, hams, &c. &c., pastry in profusion — cut in all directions — I could tell who had cut into this duck, or that goose. Some of the legs were gone — some of the wings, &c. (A good thing might be made of this.) Read the account of the banquet — the foreign ministers & many of the nobility were present. From the Guildhall, strolled through the Poultry to the Bank & New Exchange — thence, down King William Street to Fish Street Hill, & through Eastcheap to Tower Hill. Saw some fine Turkish armor (chain), every ring bearing a device. A superb cannon, cut & bored from one piece of brass — belonged to the Knights of Malta. The headman's block, upon which Kilmarnock & the Scotch lords were beheaded in the Pretender's time. The marks of the axe were very plain — like a butcher's board. — Lounged on by St. Katherine's & London Docks & Ratcliffe Highway, & within the dock walls to Wapping to the Tunnel. Crossed to Rotherhithe, & back by boat — flinging a fourpenny piece to "Poor Jack" in the mud. Took a steamer, & returned to the Temple landing & off to the Adelphi to dinner at five P.M. — dark. After dinner, Adler & I strolled over to Holborn & it being Saturday night, entertained ourselves by vagabonding through the courts & lanes, & looking in at the windows. Stopped in at a Penny Theatre — very comical — Adler afraid. To bed early.

Sunday, November 11

A beautiful autumnal day. Breakfasted about 10, & down to the Temple Church to hear the music. Saw the 10 Crusaders — those who had been to the Holy Land, with their legs crossed. Heads of the [knights], damned fine. Then down to St.Paul's — looked in a moment — then took a bus for Hampton

Court with the Doctor. Through the Strand to Piccadilly & Hyde Park, past Kensington Gardens, Kensington, Hammersmith, Chiswick, Turnham Green, Kew, & across the Thames to Richmond Hill. The royal gardens at Kew very splendid — passed the Pagoda built by Chambers. From Richmond Hill, the prospect was ineffably fine. The place is justly renowned for its beauty. The day was one in a million for England, too. Here the poet Thomson dwelt. I was on top of the coach. Pope lived near here, at Twickenham, over the way. Arrived at Hampton Court about 2 P.M. — distant some 20 miles, I think, from St. Paul's. The place is full of pictures. — Lelys & Van Dycks — the Beauties are lovely — the Dutchess of Cleveland. — Ignatius by Guido (?) — A Venus by Titian. — The Cartoons are not well disposed for light. — Oak work very rich dyed. — Rembrandt's Jew — State Beds. Walked through the parks with the Doctor till after sundown — then to a roadside inn, & drank a glass of ale. Returned to town by R.R. & dined at the Adelphi — then stopped in at St. Martin's-In-the-Fields, & to bed early.

Monday, November 12

Received another note from Mr. Bentley saying he would be in town this morning, according to my suggestion, at 12 A.M. Stopped in at the National Gallery on my way to New Burlington Street. Saw Bentley. Very polite. Gave me his note for £100 at 60 days for "Redburn." Couldn't do better, he said. He expressed much anxiety & vexation at the state of the Copyright question. Proposed my new book — "White Jacket" — to him & showed him the Table of Contents. He was much pleased with it. And notwithstanding the vexatious & uncertain state of the Copyright matter, he made me the following offer: — to pay me £200 for the first 1000 copies of the book (the privilege of publishing that number). And as we might afterward arrange, concerning subsequent editions. A liberal offer. But he could make no advance. — Left him, & called upon Mr. Murray in Albemarle Street. Not in — out of town. Strolled down St. James into the Park; took a bus & got out at the Cigar Divan on the Strand. Cheerlessly splendid. Walked to St Paul's, & sat an hour in a dozing state listening to the chaunting in the choir. Felt homesick & sentimentally unhappy. Rallied again, & down Ludgate Hill to the London Coffee House to dine according to appointment with Captain Griswold. Met Joseph Harper casually. He goes home on the Southampton. Had a noble dinner of turtle soup, pheasant &c., with glorious wine. At 10 o'clock, left with the Captain & the rest of the company (Doctor, Adler, Mulligan, McCurdy, "Stetson") for the "Judge & Jury," Bow Street. Exceedingly diverting but not superlatively moral. Nicholson is a naturally able man — so was one of the barristers. Was in bed before 1 o'clock.

According to arrangement overnight, the Doctor & I sallied out at seven o'clock (A.M.) & walked over Hungerford Bridge to Horsemonger Lane, Borough, to see the last end of the Mannings. Paid half a crown each for a stand on the roof of a house adjoining. An inimitable crowd in all the streets. Police by hundreds. Men & women fainting. The man & wife were hung side by side — still unreconciled to each other — What a change from the time they stood up to be married, together! The mob was brutish. All in all, a most wonderful, horrible, & unspeakable scene. Breakfasted about 11 A.M. & went to the Zoological Gardens, Regent's Park. Very pretty. Fine giraffes. Dreary & rainy day. Returned home at 4 P.M. & wrote up Journal. McCurdy called in about six o'clock, at the house, & bored me terribly. But I wrote a letter home meantime.

Wednesday, November 14

After a very sound night's sleep, rose much refreshed. Breakfasted at the old place in the Strand. Taylor on his last legs; some one loaned him a sovereign. But he said — "never say die." His designs upon the two ladies (awkward expression, but perfectly harmless) have failed completely. Adler & I started for the Abbey & Westminster Hall. Wandered about there awhile — went through the chapels — a knight with one wife on his right side, & a vacant space on the left — the vacant lady refusing to be put there. A serio-comico moment about Death. Up the river in a penny steamer as far as Vauxhall, passing Lambeth Palace. Began to rain hard, & arrived home drenched. Rigged up again, & in my *green* jacket called upon Mr. Murray in Albemarle St. — He was very civil — much vexed about copyright matter. — I proposed *"White Jacket"* to him — he seemed decidedly pleased — & has since sent for the proof sheets, according to agreement. Went down to the Exchange in an omnibus & tried to thrust my way in "Lloyd's" — but it was no go. Sauntered away the two hours to dinner by lounging by the shops. Opposite Somerset House in the Strand, happened to see a copy of Beaumont & Fletcher — stepped in — & found the shopman to be an old acquaintance of young Duyckinck's — Mr. Stibbs. I bought a folio of "B & F," also one of Ben Jonson. And shall purchase more. He showed me a Chapman's Homer. Dined with the Dr., McCurdy & Adler — the Dr. paying the bill. To-morrow he starts with McC. for the East — Jerusalem, &c. Went to the New Strand Theatre. A capital piece, excellently played — "The Clandestine Marriage" by Colman. All the parts admirably sustained. Mrs. Glover (an old veteran & great favorite) as Mrs. Heidleburgh. Leigh Murray as Melvil — the finest leg I ever saw on a man — a devilishly well-turned-out man, upon my soul. Old Farren was Lord Ogleby.

Special Thanksgiving day appointed by the Queen. All the shops shut as if it were Sunday, & all the churches open. Went down to the Queen's Hotel to bid good bye to McCurdy & the Dr. Thence took a bus with Adler down Newgate Street through Holborn & Oxford Street to Paddington & Edgeware Road to St. John's Wood & so round Regent's Park to Primrose Hill. The view was curious. Towards Hampstead the open country looked green, & the air was pretty clear; but cityward it was like a view of hell from Abraham's bosom. Clouds of smoke, as though you looked down from Mt. Washington in a mist. Crossed Regent's Park to New Road & got into a bus — outside — passed King's Cross & the famous Angel Inn, through City Road into Moorgate & so to the Bank. Lunched in a hole in Leadenhall Street — dismal lunch enough — & took a bus across London Bridge to the Elephant & Castle Borough. Thence walked over Westminster Bridge to the Abbey & attended service there at 3 o'clock P.M. A vast crowd present & no seats. Walked home & wrote up journal &c. At half past 5 (P.M.) went with Adler to the "Edinburgh Castle," a noted place for its fine Scotch ale, the best I ever drank. Had a glorious chop & a pancake, a pint & a half of ale, a cigar & a pipe, & talked high German metaphysics meanwhile. Home & to bed by 10 o'clock. The "Castle" is the beau ideal of a tavern — dark-walled, & like a beefsteak in color, polite waiters &c.

[32]

After breakfast at the old place in the Strand, went
to the British Museum — big arm & foot — Rosetta
stone — Nineveh sculptures — &c. From thence to
Albemarle Street — Mr. Murray was not in. Home,
& wrote to Allan by the "Canada." Walked through
Seven Dials to Oxford Street & so to Murray's again.
Found him in — was very polite, but "would not be
in his line to publish my book." Offered to give me
some of his "Hand Books" as I was going on the continent. So he sent me to the house his Book of the
Continent & for France. Met Adler by appointment
at the "Mitre Tavern," Mitre Court, Fleet Street —
the place where Dr. Johnson used to dine. His bust
by Nollekins is there. Had a "stewed rump steak." —
very fine, & bread & cheese, & ale (of course). Then
up stairs & smoked a cigar. Cosy, & comfortable
place enough. No cursed white walls. Stopped in
at the "Dr. Johnson Tavern" over the way & drank
a glass of ale. Bust of him there, also. Go to the
place through a court, where the Dr. used to live.
Some rivalry between the two places. The last, the
darkest. Thence walked down to the Queen's Hotel,
St. Martin's le Grand to see Timpson & get some
more money for the fund to purchase the book for
Capt. Griswold. P. was just off for Paris by Express.
From thence took bus for Sadler's Wells Theatre,
Islington. Pit. One of Colley Cibber's comedies —
"She would, & she would not" — Don Manuel, Mr.
A. Younge — an oldish man — seemed a favorite

with the audience — not a bad comic actor. The women (stage) were not peculiarly lovely. After the first farce (a new one, "The First of May" "by A. Younge") we walked to the Angel Inn, & took a bus across to Holborn & through Chancery Lane to Temple Bar. Stopped at the "Cock Tavern" adjoining (Tennyson's) and drank two glasses of Stout, for which the place is famous — & sells no ales. — Dark & cosy — something like the "Edinborough." Smoked a pipe also, & home with Adler & to bed.

Saturday, November 17

Adler proposed a visit to the Dulwich gallery, distant five miles, southwards. So went to the "Kings & Key" Fleet Street, and mounted a bus there. Crossed Blackfriars to Elephant & Castle — & through Newington, Walworth, Camberwell, & Denmark Hill to the beautiful hamlet of Dulwich — A most sequestered, quiet, charming spot indeed. — The gallery is full of gems — Titians, Claudes, Salvators, Murillos. — The Peasant Boys — the Venus — the Peasant Girl — Cardinal Beaufort — the mottled horse of Wouvermans — St. John — the Assumption — the old man & pipe — Mrs. Siddons as Tragic Muse. Curious old clock & tables. Met Yankee there. Left gallery, & took a little ramble round the country. Profound Calm — the green meadows & woodlands steeped in haze — strikingly English. Rode back to town on top of bus. Adler & the Yankee carrying on a spirited discussion concerning the

[34]

merits of the various paintings. — Walked towards home from the Kings & Key & concluded the affair of the present of the book to Capt. Griswold. Adler & I, however, have to pay down the most, it so happens. Upon my arrival in Craven Street, I found the letter which I expected, from Mr. Colbourn. — I should have put it down yesterday, that after leaving Mr. Murray's in the afternoon — by his instigation — I went to Mr. Colbourn's in Great Marlborough Street, was ushered up into a suite of drawing rooms & received by Mr. C.; who said, at last, I would hear from him by the next day at 3. P.M. — The letter simply declines my proposition (£200 for 1st 1000 copies) & on the ground, principally, of the cursed state of the copyright matter. Bad news enough — I shall not see Rome — I'm floored — appetite unimpaired however — so down to the Edingburgh Castle & paid my compliments to a chop. Smoked a long pipe — & then into the adjoining book store — & turned over some noble old works, & chatted with the bookseller — a very civil, intelligent young man. Bought a quarto of Davenant, & a little copy of Hudibras — Thence walked up to St. Martin's Lane, through Seven Dials — St. Giles — to Holborn & down Regent St: home to bed.

Sunday, November 18

Rose & wrote up journal till nine o'clock. Then with the Professor down to a cheap chop house near Temple Bar & had a villainous cup of coffee, a large dirty roll & a strip of fried bacon for 4 pence. Then to the Temple Church — not open — so walked the cloisters awhile. Church opened & went in & walked around a half hour. Then out & across the street to St. Dunstan's. Then down Fleet Street & Ludgate Hill (detour through White Friars) and parted with Adler at Black Friars. Took his umbrella for my cane. Went down street alone — looked in at St. Paul's — and at last entered Bow Church, Cheapside. Was shown to a seat like a pit & sat out the entire service. Curious old church indeed. Returning home, left St. Paul's Churchyard through a court towards Doctors' Commons & wandered among a labyrinth of blind alleys & courts to Apothecaries' Hall & Printing House Square. Then crossed Fleet Street near Temple Bar, towards the North & threaded more alleys, lanes, & courts, & so home at 3 P.M. of a dismal drizzling day. Read a little about the Mannings murder — guide-books &c — wrote Lizzie. Dined with Adler at the Adelphi — then to the Rainbow Tavern (Tennyson's) & smoked a cigar in a room inhabited by a melancholy man. Splendid dining room at the Rainbow. Terrific bumper of Stout. Famous for Stout. Walked off alone down Fleet Street & went to St. Bride's Church. Woman showed me to a big pew — almost unasked — blushed a good deal, what with Stout, jam, heat, & modesty. Excited vast deal of gazing somehow. Good sermon — a charity one. Gave sixpence. Home & to bed.

Rose & wrote. After breakfast with Adler went down to the city & presented my letter to Thomas Delf (Duyckinck's). Not in. Then to David Davidson (Young D's) Paternoster Row — Wiley's Agent. Not in. Then to J. M. Langford P. Row (Mrs. Welford's). Very civil reception — invited me to go to hear McCready same night — also to sup with him & meet Albert Smith, the comic writer who has just returned from the East & purposes writing something "funny" about it. Leaving Mr. Langford, went to the Longmans in the Row & proposed "the book." Saw Mr. Longman himself — very polite — promised to send me a note by evening. Took a bus up to Mr. Bentley's to see if there should be any letters. Found two budgets from home — much delighted me — also other letters. On first entering the place, a "clark" steps up — "Lord John Manners has been here for you, sir!" The devil he has — "And left this letter, Sir." — Shove it along then. His Lordship's note was very kind indeed. He enclosed two letters to Monckton Milnes & Lady E. Drummond. Went into B's private room & read my home letter. All well, thank God — & Barney a bouncer. Went to Mr. Murray's. Told him, that people here having anticipated me, I should stay awhile now, & make some social calls, &c. He took me upstairs to see his gallery of literary portraits. Fine head of Byron, Moore, Campbell, Borrow &c. &c. (Among other letters from the U.S. I received one from Willis, enclosing a letter to Lord J. Manners & M. F. Tupper. The letter to L. J. Manners is from his sister in New York.) Went home & answered Manners' note, saying I would call tomorrow. Having no one to send it by, took it myself to the Albany, & handing it to the beadle to be delivered, was told that Lord J. M. had that morning left town. Wrote him therefore at Belvoir Castle. Also sent off letter to Lizzy by the *Herman.* Dined with Adler at the Adelphi. Then home. Mr. Langford called, & bidding poor Adler goodbye (he is off for Paris at midnight by the steamer), went to the Haymarket. Full house. Went into the critics' box, "Times" & "Herald" men there. Macready panted hideously. Didn't like him very much upon the whole — bad voice, it seemed. James Wallack, Iago, very good. Miss Reynolds, Desdemona — very pretty. Horrible Roderigo. Stayed out the first farce — "Alarming Sacrifice" by Buckstone the great commedian, who played the principal part — Bob Ticket. Very funny. Home & to bed at 12. Found a note from Murray inviting me to dine & meet Lockhart on Friday. Also note from Capt. Griswold acknowledging volume.

Tuesday, November 20

Opened the door & found a note on top of my blackened boots. It was from the Longmans, saying they abided by their original terms. Breakfasted alone at the old place — capital coffee, that — a nice neat place indeed & the morning papers delightfully *fresh*. The old lady too very nice. Returned home & read & wrote till 12 o'clock. Then got out some of my letters & resolved to go at it, like a regular job, which it is, this presenting letters of introduction. First went to Mr. Rogers in St. James's Place — very quiet & deathlike — not in — left letter & card. Then across the Park to Upper Belgrave Street to the American minister. Found both him & Davis the Secretary. Mr. Lawrence received me very kindly, said that the Duke of Rutland (father of L. J. Manners & L. Emmeline) had been seeking my address from him. The D. left town yesterday. Mr. L. invited me to call same evening at the Clarendon Hotel to see his wife &c. Learning from him that Lady E. Drummond was in town I went home, rigged up & jumped into an omnibus for Oxford Street. Walked from thence to Bryanstone Square — out of town — wouldn't return till after Christmas. Left letter and card. Then walked to Upper Harley Street (devil of a walk, too) and left letter & card for Russell Sturgis. Also walked to Portman Square & left same for Joshua Bates. Then walked down street, stopped & learnt the address of Moxon the Publisher. Found him in — sitting alone in a back room — he was at first very stiff, cold, clammy, & clumsy. Managed to bring him to, though, by clever speeches. Talked of Charles Lamb — he warmed up & ended by saying he would send me a copy of his works. He said he had often put Lamb to bed —

drunk. He spoke of Dana — he published D.'s book here. Thence, down to the Adelphi by bus & dined alone — fried sole. Then home, & read & wrote. At 7 P.M. called at the Clarendon Hotel, Bond Street, upon Mr. & Mrs. Lawrence. Found them in a noble drawing room. Mr. Lawrence was very kind, unaffected & agreeable. I like him much. He is a very fine-looking benevolent-seeming man. But his wife! Such a sour, scrawny scare [?] scum was never seen till she first saw herself in the glass. I do not fly out at her for her person — no, but her whole air & manner — God deliver me from such horrors as Mrs. Lawrence possesses for me. Her skinny, scrawny arms were bare — She talked of Lady Bulwer — said that — but there is no telling how she managed so well her veiled & disgusted air, without being at all uncivil or meaning any incivility. She belongs to that category of the female sex there are no words to express my abhorrence of. I hate her not — I only class her among the persons made of reptiles & crawling things.

Coming home stopped at a place in Haymarket — singular interview there for a moment. Home early & to bed by 9½.

Wednesday, November 21

Breakfasted & took a ha'penny steamer at the Adelphi and down to London bridge — thence another steamer for Greenwich Hospital. Crowds of pensioners. Went into the Painted Hall — sea pieces &

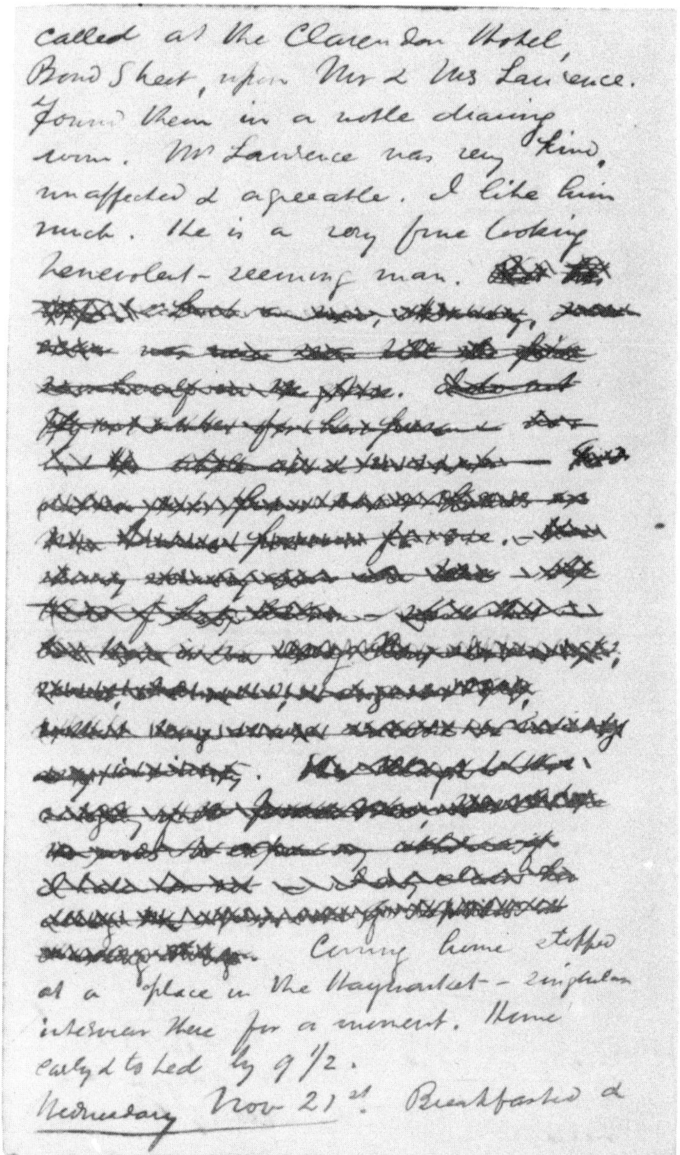
Called at the Clarendon Hotel,
Bond Street, upon Mr & Mrs Lawrence.
Found them in a noble drawing
room. Mr Lawrence was very kind,
unaffected & agreeable. I like him
much. He is a very fine looking
benevolent-seeming man. [redacted] Coming home stopped
at a place in the Haymarket — singular
interview there for a moment. Home
early & to bed by 9 1/2.
Wednesday Nov 21st. Breakfasted &

Facsimile — see page 40

portraits of naval heroes, coats of Nelson in glass cases. Fine ceiling. Was shown about the wards by a machine. Visited the Chapel. Fine painting by West of St. Paul. Saw the Pensioners at dinner. Over 1500. Remarkable sight. The negro. — Hat off! — Hat on! — Married men & unmarried — mess apart. 2 & 2. Pensioners in palaces! Story of Charles II. Walked in Greenwich Park. Observatory. Fine view from a hill — talk with an old pensioner there. Home by railway. From London Bridge walked to Delf's — not in. Thence to Davidson's in Paternoster Row. Had a talk with him — invited me to dine — went to Queen's Hotel. (Previously I stopped in at the Longmans about the book.) Dined on oxtail soup — chops — ale — port wine. Good dinner & David a good fellow. Strolled through Fleet Market — butchering under hatches — blubber rooms. Walked with Davidson to American Bowling Saloon. Made me go in. Rolled one game & beat him. Then home & dressed for Mr. Langford's. On the way bought Lavater's Physiognomy in Holborn for 10 shillings (sterling). Found Langford & a young fellow in his lodgings. Snug place enough. At last there came in four or five young fellows — sociable chaps. And in the end, Albert Smith, the comic writer, & Tom Taylor, the Punch man & Punch poet. Smith was just from the East, & sported a blazing beard. A rattling, guffaw cockney — full of fun & a little malice perhaps. Nice plain supper — no stiffness. Porter passed round in tankards. Round table.

Potatoes in a napkin. Afterwards, Gin, brandy, whiskey & cigars. Smith told funny stories about his adventures in the East, &c. Gave me his address, &c. Came away about 2 A.M. & through Oxford Street home & turned flukes. Found Moxon had sent me his copies of Charles Lamb . . . . also found a card from Russel Sturgis & a note from Joshua Bates inviting me to dine & spend Sunday with him at East Sheen. Accepted.

Thursday, November 22

Rose late — headache — breakfasted & off on top of omnibus to Great Western Railway for Windsor. Had to wait an hour. Pleasant day. Round Tower — fine view. Long Walk. Went through the state apartments. Cheerlessly damnatory fine. Mast of the Victory & bust of Nelson. Shield of Cellini. Gobelin tapestry is miraculous. Made the acquaintance of an Englishman — viewed the royal stables with him. On the way down from the tower, met the Queen coming from visiting the sick Queen Dowager. Carriage & four, going post with outriders. The Prince with her. My English friend bowed & so did I — salute returned by the Q., not by the P. I would commend to the Q. Rowland's Kalydore for clarifying the complexion. She is an amiable domestic woman though, I doubt not, & God bless her, say I, & long live the "prince of whales." The Stables were splendid, Endless carriages &c. Walked in the great Park with Englishman talking about America.

Arrived in London about 5 P.M. Dined & home. Found my book (Lavater & ½ sovereign, on top of it) in my room. Also note from Mr. & Mrs. Lawrence inviting me to dinner, which, had to decline owing to prior engagement with Mr. Murray. Went & left my reply in person at the Clarendon Hotel. Then to Bentley's — not in. Home & wrote up journal.

Friday, November 23

A long night's rest — having turned in at 9½ & rising at about the same hour this morning. Must have caught cold last evening on top of the omnibus from R. R. Station. Felt feverish & chilly by turns. Light breakfast, & down the Strand. Called on Capt. Griswold — not in. Then to the Aldine Chambers, & saw Davidson. Took me upstairs, & had a chat. He said he thought he could get my note (Bentley's) cashed for me — by which, felt much relieved. Talked to him concerning White Jacket & copyright. He said I must keep pushing & mentioned the names of some more publishers whom I ought to try. At his instance, he went with me to Bogue's (Strand) & introduced me. I stated my business. B. was all ears. Gave him an idea of the book. He finally said he would send me his answer by six o'clock this evening. It is now 3½ P.M.; & I predict that he declines. We shall see. After leaving Bogue, walked in the Temple courts & gardens. Then, went musing along through Drury Lane to Holborn, & came home with

[43]

a fit of the Blues. On my way, bought a pound & a half of figs & brought them home to make lunches of. Wrote Lizzie & Allan by the Steamer which goes to Boston this time.

¼ to 11 P.M. I have just returned from Mr. Murray's where I dined agreeably to invitation. It was a most amusing affair. Mr. Murray was there in a short vest & dress-coat, looking quizzical enough — his footman was there also, habited in small clothes & breeches, revealing a despicable pair of sheepshanks. The impudence of the fellow in showing his legs — & such a pair of legs too! — in public, I thought extraordinary. The ladies should have blushed, one would have thought — but they did not. Lockhart was there also, in a prodigious white cravat, (made from Walter Scott's shroud, I suppose). He stalked about like a half galvanized ghost — gave me the tips of two skinny fingers when introduced to me — or rather, I to him. Then, there was a round-faced chap by the name of Cooke — who seemed to be Murray's factotum. His duty consisted in pointing out the portraits on the wall, & saying that this or that one was esteemed a good likness by the high & mighty ghost Lockhart. There were four or five others present — nameless, fifth-rate-looking varlets — & four lean women, one of them agreeable in the end. She had resided some time in China — I talked with her some time. Besides these, there was a duodecimo footman, or boy in a tight jacket with bell-buttons. — At dinner the stiffness, formality, & coldness of the party was wonderful. I felt like knocking all their heads together. I managed to get through with it, however, somehow by conversing with Dr. Holland, a very eminent physician, it seems, & a very affable, intelligent man, who also has travelled immensely. After the ladies withdrew, the three decanters — Port, Sherry & Claret — were kept going the rounds with great regularity. I sat next to Lockhart, and seeing that he was a customer who was full of himself & expected great homage; & knowing him to be a thorough-going Tory & fish-blooded Churchman & conservative, & withal, editor of the Quarterly — I refrained from playing the snob to him like the rest — & the consequence was he grinned at me his ghastly smiles. After returning to the drawing room, coffee & tea were served, & I soon after came away. Such is a publisher's dinner — a comical volume might be written upon it. Oh Conventionalism, what a ninny, thou art, to be sure. And now, I must turn in. But first, let me add, that upon coming home, I found a note for me from Bogue, the publisher. I knew its contents at once — there seemed little use in opening it. He declined; alleging among other reasons, the state of the copyright question. — So we go.

Saturday, November 24

Upon sallying out this morning encountered the oldfashioned pea soup London fog — of a gamboge color. It was lifted, however, from the ground &

floated in mid air. When lower, it is worse. Lamps lighted, as it was in the old lady's room where I took my usual breakfast — two small rolls, a bit of butter the size of a dollar, & about the thickness, — & two cups of incomparable coffee. After digesting the "Times" & the rolls, went down street & stopped in at Chapman & Hall's the publishers. Saw Mr. Chapman, a very gentlemanly man. Proposed the book. He at once threw the copyright matter at me. No go. Then went to the Aldine Chambers & saw Davidson who very kindly said he would get my note cashed. Left it with him, & went to H. G. Bohn's, York Street, Covent Garden. A florid fellow in a crimson vest. Proposed the book. No go. Walked through the market, home.

Midnight. Just returned from East Sheen with an indefinite quantity of Champagne, Sherry, Old Port, Hock, Madeira, & Claret in me. But first to bring down my chronicle. After seeing Bohn (as above) & coming home, I was called upon by Mr. Stevens, who travelled all the way up to my room here in the 4th story. He proves to be a very fine fellow, and I was the more pleased with him, & in some sort loved him, from the circumstance that he told me [he] had been acquainted with "my wife," that is to say with Dolly. He stayed so long, however, that little time was left me to dress for Mr. Bates'. Being rigged at last, though, I sallied out, jumped into a bus & down to St. Paul's — there, took a Richmond stage, & away to East Sheen — 9 miles. The bus was crowded.

Alighted in a dark foggy lane, & picked my way towards a distant illumination which proved Mr. Bates'. Upon entering the vestibule, some dozen footmen in small clothes & gold lace received me, and I was ushered into a sumptuous chamber up stairs. Descending into a parlor, Mr. Bates gave me a courteous welcome. Shortly after, a mustache entered the room & sat down in a chair opposite — a Vienna man. The company gradually all appeared, & we went into dinner. On my left was a nephew of Lord Ashburton, & the Baroness de Somebody I don't know who. On my right was Mr. Peabody, an American for many years resident in London, a merchant, & a very fine old fellow of fifty or thereabouts. There was a Baron opposite me, and a most lovely young girl, a daughter of Captain Chamier the sea novelist. Half the company were foreigners. The dinner was superb — the table was circular — the service very rich. I ate of sundry mysterious French dishes. Everybody was free, easy & in good humor — all talkative & well-bred — a strong contrast to the miserable stiffness, reserve, & absurd formality of Mr. Murray's, the tradesman's, dinner last night. Mrs. Bates is a fat, ugly, good-natured, amiable woman of sixty. The Baroness is a pretty, vivacious woman of thirty. Her husband, a round, sleek man, who said little, but ate his dinner with a relish. Mr. Russel Sturgis was there. Somehow he reminded me of Murray's man, Cook, last night. Mrs. Sturgis is a fine-looking lady. The wine went round freely

[47]

after the ladies withdrew. Upon entering the drawing room, coffee & tea were served. The house is a
large & noble one — the rooms immense — the deco-
rations brilliant — statuary, vases, & all sorts of costly
ornaments. I saw a copy of *Typee* on a table. Mr.
Bates seemed to be quite a jolly old blade. I had intended to remain over night agreeably to invitation,
but Peabody inviting me to accompany him to
town in his carriage, I went with him, along with
Davis, the Secretary of Legation. — By the way,
Mr. Stevens invites me to dinner tomorrow (Sunday) at Morley's. — Mr. Peabody was well acquainted
with Gansevoort when he was here. He saw him
not long before his end. He told me that Gansevoort
rather shunned society when here. He spoke of him
with much feeling. No doubt, two years ago, or
three, Gansevoort was writing here in London, about
the same hour as this — alone in his chamber, in profound silence — as I am now. This silence is a strange
thing. No wonder the old Greeks deemed it the vestibule to the higher mysteries.

Sunday, November 25

Passed a most extraordinary night — one continuous
nightmare till daylight. Hereafter if I should be
condemned to Purgatory, I shall plead the night of
November 25th 1849 in extenuation of the sentence.
I impute the nightmare to a cup of prodigiously
strong coffee & another of tea, which I took at Mr.
Bates' just previous to leaving. At daylight turned

[48]

over & had a nap till about 10 o'clock. Dressed and emerged into a fog — quite thick. As my old coffee haunt in the Strand is closed on Sundays, I steered my way up St. Martin's Lane, intending to breakfast at the Hotel de —— commended to me by George Duyckinck, as "good & cheap." Lost my way in the fog, & stopped at an atrocious hole where I got coffee & roll for the sum of fourpence — including a thimbleful of sugar. Drunken fellow there. Came at last upon Leicester Square, & as my breakfast had been most meagre & mean, proposed to myself a cup of French Coffee at the Hotel de —— Stepped in & called for it, & very good it was. Place full of Frenchmen. Charged me 18 pence for the coffee & small roll. — Thence to the Abbey & attended service. Abbey full of fog. Thence walked in St. James's Park. Saw the sentries (mounted) at the Horse Guards relieved. Thence down to Edinborough Castle and read the Sunday Times & Despatch. Thence down to Queen's Hotel to see Davidson. Not in. Thence down street further yet, & wandered about Threadneedle, Throgmorton, Leadenhall Streets, &c., passed Aldgate Pump into White Chapel Road (where Bayard Taylor stopped). Passed Moses & Son establishment. Thence took bus to St. Paul's to service. Thence by bus home at 5 P.M. And thus closes a most foggy, melancholy, sepulchral day. But in half an hour's time from now I go to Morley's to dine with Stevens & Davis — & there I hope to recover myself in the companionship

[49]

& conversation of mortals. — "Oh Solitude! where are the charms" &c. —

Tuesday, November 27

Dined at Morley's & had a very pleasant time, on Sunday. Mr. Davis and a Mr. Newton (chaplain in the Navy) were there. Newton is a fair sample of an idle man. He seems a good-natured, well disposed sort of a liver upon Uncle Sam, however. — No more. I am on board the steamer "Emerald" bound for Boulogne — we are just in the Nore, and the jar & motion are so great that writing is too hard work. I must defer bringing up my journal till I reach terra firma. After the lapse of a few days, I find myself this Thursday night, snugly *roomed* in the fifth story of a lodging house No. 12 & 14 Rue de Bussy, Paris. It is the first night that I have taken possession, & the "Bonne" or chambermaid has lighted a fire of wood, lit the candle, & left me alone, at 11 o'clock P.M. On first gazing around, I am struck by the apparition of a bottle containing a dark fluid, a glass, a decanter of water, & a paper package of sugar (loaf) with a glass basin next to it. — I protest all this was not in the bond. — But though if I use these things they will doubtless be charged to me, yet let us be charitable, — so I ascribe all this to the benevolence of Madame Capelle, my most polite, pleasant, and Frenchified landlady below. I shall try the brandy before writing more. — And now to resume my Journal. To go back to Sunday afternoon, November 25th. After dining with Mr. Stevens agreeably to invitation at Morley's, I smoked a cigar with Davis, & then we went to the Clarendon Hotel & called upon Mrs. Lawrence. Found the lady & the Minister in. Mrs Lawrence was so very pleasant this evening that I must take back the bitter things I said of her before.

They have taken the Earl Cadogan's house in Piccadilly, to reside in. Coming home with Davis I was struck with his expressions concerning the poverty & misery of so large a portion of the London population. He revealed a heart. Next day — Monday, November 26 — after breakfast went to Davidson's to see about my note's getting cashed. Waited some time. At last he came in, gave me a cheque on his bankers, & off I went to Livingston & Wells down Lombard Street, & deposited £40 for Allan to draw in New York. When I went to the bankers, they shoveled the sovereigns over to me in curious style. At 2 P.M. called for Stevens at Morley's, & went with him & Newton to the library of the British Museum. Endless galleries & three-deckers of books. Saw many rareties. — Maps of London (before & after the Great Fire), Magna Charter — Charlemagne's bible — Shakespeare's autograph (in Montaigne), &c. &c. &c. Went into the Manuscript room — saw the famous Alexandrian Manuscript & many Saxon Mss. of great value. Comical librarian. — Dined at the "Mitre," Fleet Street, with Davidson, & paid the joint bill myself. Then to the "Blue Posts," Cork

Street, & took some of their renowned punch. Wrote Allan from thence. Then home & to bed, after packing my little portmanteau, which I had bought in London for traveling on the continent. On Tuesday — the 27th November — rose early, paid my bill — 3 guineas — & into a cab, & down to London Bridge stairs for the Boulogne boat. Took a fore cabin passage (8 shillings — saloon 12). Fine run down the river, & fine cold morning. Passed Gravesend, Tilbury Fort (midway to the sea); passed Sheerness near the mouth, & Margate. Great number of ships running out. Passed by the Goodwin Sands, Deal, South Foreland, Dover, & then across the Channel to Boulogne. A shocking accident occurred to one of the hands of the boat. His foot was ruined in the machinery. All the passengers seasick, but three or four. Young Frenchman consoling them. Arrived in Boulogne about 8½ P.M. Marched off to the Custom House & surrendered Passport. Then to the Bedford Hotel & took some tea & cold beef. Then to a cold bed. Rose early & took the 7 o'clock train for Paris, — distant 160 miles or more. Took "third class" train (fare about 15 francs). Fine day & rode through a charming country. Did not stop at Amiens to see the Cathedral there — not time. Arrived in Paris (Rue Lazare) at 4 P.M. Took a cab with the young Frenchman & off to No. 3 Rue de la Convention to find Adler. Could hear nothing of him. So went to "Meurice's," Rue de Rivoli & took a room. Dined at table d'hôte at a little after 5 P.M. Splendid table — French dishes — ate I know not what. Talked with two Englishmen at table. After dinner went to Galignani's — Adler's address not there.

Subscribed to the reading room. Then walked to the Palais Royal. Gorgeous cafés there & shops — capital of Paris. Turned in early.

Thursday, November 29

Today rose early, breakfasted at a café in the Rue de Rivoli & walked to the Place de la Concorde — magnificent. Crossed the Seine towards the Chamber of Deputies. Returned & met great numbers of troops marching all about. Like a garrisoned town. Wrote two letters to Adler. Went to the "Latin Quarter" in search of Madame Capelle — found her, engaged a room & recrossed the river. Visited the Bourse — great hum round a mystic circle — & Livingston & Wells's. Dined in the Rue Vivienne for two francs — several courses including a bottle of Bordeaux. Then went to "Meurice's" — paid my bill (15 francs for a dinner, a bed & a sheet of paper) and drove to my present quarters on the other side of the Seine. Then walked to the Palais Royal (National now) & went into the pit of a theater there (Admittance 25 sous) Three comical comedies — plenty of babies & wet napkins &c. Came away early & home — smoked & to bed.

Next day Friday, breakfasted over the way at the "Café Francais." Very good place. Then down street with a blue nose (so cold it was) & visited Notre Dame. They are repairing it. A noble old pile. Then walked about the old city & crossed to Rue St. Antoine, to the Place de la Bastille, where now stands the column of July. Café la Bastille & villain's hole there. Then in a bus (roundabout way) to the Louvre, & spent three hours in the Museum. Heaps of treasures of art of all sorts. Admirable collection of antique statuary. Beats the British Museum. Then to the Bourse & dined at the "Rosbif" Restaurant. Then to Galignani's & read the papers. Late arrival from New York. Great excitement again about California. Saw that the thing called "Redburn" had just been published. Extract from "Lit. World" notice &c. Having received a note from Adler (to my great joy) who said he would be at his rooms at ¼ past 7 this evening, I called accordingly. — Not in — so sat & jabbered as well as I could with Madame till he arrived. Was rejoiced to see him — we went together to his room — he brought out tobacco & we related our mutual experiences. Left him at nine o'clock & came home. Fire made & try to be comfortable. But this is not home & — but no repinings.

Coffee & roll at 10½ at Café over the way. At 11 A.M.
Adler called & we started for the Hôtel de Cluny.
Stumbled upon the Sorbonne, & entered the court.
Saw notices of Lectures — Cousin, &c. — Hôtel de
Cluny proved closed. Took bus & went to Père le
Chaise. Fine monument of Abélard & Héloïse.
Tombs of generals, &c. Returning, visited Abattoir
of Popincourt. Noteworthy place, founded by Na-
poleon. Old woman, &c. Thence to the Boulevards
& through them to Rue Vivienne to the Bourse &
Livingston & Wells'. Dined with Adler at the "Ros-
bif" & thence to his room in Rue de la Convention.
At 7½ P.M. left for the Palais Royal to see Rachel in
Phèdre. Formed in the "cue" in the arcade — stood
an hour or two & were cut off. Returned bitterly disappointed to Adler's room. Bought a bottle of Bordeaux (price 8 sous) & drank half a glass of it. Home
at 11 & to bed.

Rose about 10, & walked over to Palais Royal, where
breakfasted at Café d'Orléans. Then to Adler's
rooms & with him to the church of St. Roch — noble
interior — ladies' chapel — paintings &c. Singularly
eloquent preacher — funeral. Thence to the Made-
leine — went in. Superb exterior. Thence through
Place Vendôme, across Place de la Concorde over the
river to Hôtel des Invalides. In the Chapel saw the
Austerlitz flags, &c. Algiers, &c. Dining rooms. Sil-

ver plate of officers. Gallery of paintings. Thence through out-of-the way streets to the Luxemburg. Fine gallery of modern French school. Large park, gardens. Chamber of Peers. Thence to the Sorbonne. Got into the church, & saw tomb or rather monument of Cardinal Richelieu. Good Sculpture. Adler here left me. Went to the Panthéon — still building. Thence into an old church nearby, & through Middle Age courts & lanes to the river & through Place du Carrousel, to Adler's. At 5 P.M. dined at his table d'hôte. Two members of the Chamber there. Several Yankees. Any quantity of Bordeaux. Cold room & chilly wine. A lady talking about *"Flemish"* things. After dinner Hotchkiss (A's friend) joined us in A's room; & we had a talk, with some Eau-de-vie & cigars. At 10 sallied out into a dark, rainy night, & made my melancholy way across the Pont (within a biscuit's toss of the Morgue) to my sixth story apartment. I don't like that mystic door (tapestry) leading out of the closet.

Monday, December 3

Went to Galignani's after breakfast to get Continental Guide. Returning met Adler in the Rue de la Fontaine Molière. Thence walked through Champs Élysée to the Triumphal Arc de l'Étoile. Ascended it — splendid view of Paris. Bought little medal of old woman. Took bus to the Pont Neuf. Got my original passport at the Police. Looked in at the Morgue. Dined at the "Rosbif." Called on Adler about five (P.M.), smoked a cigar till his return from dinner. Then, went with him to the "Opéra Comique" Boulevard des Italiens. Two pieces, house crowded. Splendid orchestra. Home & to bed, after stopping with Adler at the Café de Foy for a cup of "chocolat," which was exquisite.

Tuesday, December 4

Met Adler at the "Bibliothèque Royale" (Cardinal Mazarin's). Looked over plates of Albert Dürer, & Holbein. Walked through the halls of books. Saw autographs, Persian MSS, &c. Coptic, &c. Franklin's letter. Parnassus, Museum. Left Adler at his room & walked to the American Minister about passport. Thence to the Prefecture of Police, &c. Devilish nonsense the whole of it. Bought two pair of gloves, one pair of shoes for Lizzie. Bought a copy of *Telemachus* for self near the Louvre. Called on Adler at his room & invited him to dinner. Went to the Italian Boulevard and dined at the Café Anglais. Thence to Café de Cardinal for chocolat. Thence to his room and talked high German metaphysics till ten o'clock. Then home & wrote up journal, as above, and now to bed.

Wednesday, December 5

Breakfasted at a cheap place in the Place du Carrousel. Wandered about the stalls along the quays & bridges. Went to the Museum Dupuytren. Pathological. Rows of cracked skulls. Skeletons & things

[57]

without a name. Thence to the Hôtel de Cluny. A most unique collection. The house is just the house I should like to live in. Glorious old cabinets — ebony, ivory carving. — Beautiful chapel. Tapestry, old keys. Leda & the Swan. Descended into the vaults of the old Roman palace of Thermes. Baths, &c. Thence lounged along the quays to the Palace of the Deputies. Sauntered in past the guards. And would have gained admittance to the Chamber but for the unlucky absence of my passport, which I had left to get the "visas" of the Belgian & Prussian ministers. Thence to Adler's — He invited me to dine with him. Went to the No 3 Rue de la Convention and dined with three Americans. Thence to the Théâtre Francais, to hear Madame Rachel. Formed in the *"queue"* but gave it up at last. Back to Adler's, smoked a cigar & crossed the river home & to bed early.

Thursday, December 6

Took my last breakfast in Paris — at the Café Francais, Rue de Bussy. Thence to Galignani's & read the papers. Thence took bus for the R. R. Depot & off to Versailles. Old man for guide through the gardens. A most magnificent & incredible affair altogether. Splendid paintings of battles. Grand suite of rooms of Louis Le Grand. Titan overthrown by thunderbolts, &c. Apollo & the horses on the fountain. Back to Paris by dusk. Took a cab & went to Meurice's & got my passport. Thence to my rooms in

Rue de Bussy — paid my bill, &c. — took a last adieu of Madame Capelle & went to Adler's. Dined at a Restaurant near Palais Royale & with Adler went to his room — where I now write up my last Parisian journal. — Selah!

Brussels

Sat up conversing with Adler till pretty late, — (Topic — as usual — metaphysics.) Then turned in, in a room below him with orders to be roused at 6½ A.M. Rose accordingly & dressed in the dark — taking my usual bath. Off to the R. R. Station in a cab. Fare to Brussels something like 27 francs — 2d class cars. A dull, dreary ride all day over an interminable flat pancake of a country. Passed through Amiens (second time: couldn't see the cathedral) — Arras (where were an immense number of windmills), & several fortified towns — Douai, Valenciennes (renowned for lace). At Quièvrain passports were demanded. At Valenciennes we entered the Netherlands or Belgium — Low Countries indeed. — And so through Mons & Soignies to Brussels. Drove a prodigious way up a long narrow ravine of a street to the Hôtel de France, where I arrived about 7½ P.M. in the dark. Was accompanied by a fellow passenger from the R. R. — a phlegmatic impracticable North of England man. He conversed, however, a little upon America. Took tea with him — then a short stroll &

here I am in a fine chamber, all alone, in the capital of Belgium.

There was a sound of revelry by night,
And Belgium's capital had gathered then her beauty
*   and her chivalry . . .*

But no sound of "revelry" now, heaven knows. A more dull, humdrum place I never saw — though it seems a fine built place. Waterloo is some 8 miles off. Cannot visit it — & care not about it. Tomorrow I am off for Cologne. It is a chance whether I can get up the Rhine or not — owing to the ice. Must try it though.

Saturday, December 8

Breakfasted *tête-à-tête* with the English phlegmatic — he said nothing & I ditto. Having determined to travel into Germany entirely unencumbered, I left my carpet-bag — or rather, portmanteau — at the hotel, and after a short stroll (during which I walked through the Palace Park & by the Palace, & past the Botanic Gardens of the King, & along the miniature Boulevards) I went to the R. R. Station & took my seat for Cologne. I took a second class car, which was as pleasant as the 1st class in America & incomparably better than the corresponding class in England. For fellow "voyagers" I had a cocked-hatted priest & several dark-eyed young women in crape hoods. One of them looked mysterious. For the first 60 miles or so our way lay through the same flat, dead country, as from Paris to Brussels. But at Liége (the gun place) the scenery changed. It was all hill & valley & we wound through the most romantic defiles imaginable — by old ruined mills & farmhouses of the Middle Ages. At the Prussian frontier my passport was demanded & taken. In the car was a young Frenchman — a genuine blade, who carried a flask of Belgian Gin in his pocket with a glass, & he frequently invited me to drink. He had been in America. At Aix-la-Chapelle he & party left us, & there only remained in the car with me, a young Berlin man, who talked a little English. About 9 o'clock we arrived at Cologne, in the dark, and taking a "bus" drove to a Hotel — I supposed that it was the Hôtel de Cologne, having been advised to go thither by the Guide Book. The place proved to be a great, cold, vacant palace (half stable). I took tea (à la Français) and retired early into a German bed, with a pillow at my feet. I intended taking the boat at 10¼ in the morning, & so slept sweetly dreaming of the Rhine.

Sunday, December 9

Cologne

Sallied out before breakfast and found my way to the famous cathedral, where the everlasting "crane" stands on the Tower. While inside was accosted by a polite worthy who was very civil pointing out the "curios." He proved a "valet de place." He tormented me home to the Hotel & got a franc out of me. Upon going to the Steamer Office I learned that no boat would leave that morning. So I had to spend the day in Cologne. But it was not altogether unpleasant for me so to do. In this antiquated gable-ended old town — full of Middle Age, Charlemagne associations — where Rubens was born & Mary De Medici died — there is much to interest a pondering man like me. But now to tell how at last I found that I had not put up at the "Hôtel de Cologne," but at the "Hôtel du Rhin" — where my bill for a bed, a tea, & a breakfast amounted to some $2, in their unknowable German currency. Having learnt about the Steamer, I went to the veritable Hôtel de Cologne (on the river) & there engaged the services of a *valet de place* to show me the sights of the town for 2 francs. We went to the Cathedral, during service — saw the tomb of the *Three Kings of Cologne* — their skulls. The choir of the church is splendid. The structure itself is one of the most singular in the world. One transept is nearly complete — in new stone, and strangely contrasts with the ruinous condition of the vast unfinished tower on one side. From the Cathedral we went to the Jesuits' Church, where service was being performed. Thence to the Museum & saw some odd old paintings; & one splendid one (a sinking ship, with the Captain at the mast-head — defying his foe) by Scheffer (?). Thence to St Peter's Church & saw the celebrated *Descent from the Cross* by Rubens. Paid 2 francs to see the original picture turned round by the Sacristan. Thence home. Went into a book store & purchased some books (Views & Panoramas of the Rhine) & then to the Hotel. At one o'clock dinner was served (Table d'hôte), a regular German dinner & a good one, "I tell you." Innumerable courses — & an apple pudding was served between the courses of meat & poultry. I drank some yellow Rhenish wine which was capital, looking out on the storied Rhine as I dined. After dinner sallied out & roamed about the town — going into churches, buying cigar of pretty cigar girls, & stopping people in the street to light my cigar. I drank in the very vital spirit & soul of old Charlemagne, as I turned the quaint old corners of this quaint old town. Crossed the bridge of boats, & visited the fortifications on the thither side. At dusk stopped at a beer shop — & took a glass of *black ale* in a comical flagon of glass. Then home. And here I am writing up my journal for the last two days. At nine o'clock (3 hours from now) I start for Coblenz — 60 miles from hence. I feel homesick to be sure — being all alone with not a soul to talk to — but then the Rhine is before me, & I must on. The sky is overcast, but it harmonizes with the spirit of the place.

Monday, December 10

Coblenz.

Embarked last night about 9½ P.M. for Coblenz. But before so doing went out after tea to take a final stroll through old Cologne. Upon returning to the hotel, found a large party assembled, filling up all the tables in the Dining Saloon. Every man had his bottle of Rhenish, and his cigar. It was a curious scene. I took the tall spires of glasses for castles & towers, and fancied the Rhine flowed between. I drank a bottle of *Rüdesheimer* (?). When the boat pushed off it was very dark, & I made my way into the 2d cabin. There I encountered a German who was just from St. Louis in Missouri. I had a talk with him. From 9½ P.M. till 5 A.M. I laid down & got up, shivering by turns with the cold. Thrice I went on deck, & found the boat gliding between tall black cliffs & crags. — A grand sight. At last arrived at Coblenz in the dark, & got into a bed at the "*Géant Hoff*" near the quay. At ten o'clock in the morning descended to breakfast, & after that took a *valet de place* & crossed the Bridge of Boats to the famous Quebec fortress Ehrenbreitstein. A magnificent object truly. The view from the summit is superb. Far away winds the Rhine between its castellated mountains. Crossed the river again, & walked about the town, entering the curious old churches, half Gothic, half Italian — and crossed the Moselle at the stone bridge near where Prince Metternich was born. Singular that he was born so near the great fortress of Germany. Still more curious that the finest wine of all the Rhine is grown right under the guns of Ehrenbreitstein. At one o'clock dined at "The Géant" at the table d'hôte. There were some six or eight English present — two or three ladies & many German officers. The dinner was very similar to the dinner at the Hôtel de Cologne yesterday. After dinner walked out to the lower walls & into the country along the battlements. The town is walled entirely. At dinner I drank nothing but Moselle wine — thus keeping the counsel of the "Governor of Coney Island" whose maxim it is, "to drink the wine of the country in which you may be travelling." Thus at Cologne on the banks of the Rhine, & looking at the river through the window opposite me — what could I imbibe but Rhenish? And *now,* at Coblenz — at the precise junction of the Moselle — what regale myself with but Moselle? — The wine is bluish — at least *tinged* with blue — and seems a part of the river after which it is called. At dusk I found myself standing in the silence at the point where the two storied old rivers meet. Opposite was the frowning fortress — & some 4000 miles was America & Lizzie. Tomorrow I am *homeward-bound!* Hurrah & three cheers!

I last wrote in my Journal on the banks of the Rhine — & now after the lapse of a few days I resume it on the banks of the Thames, in my old chamber that overlooks it, on Saturday the 15th of December, '49.

I broke off at Coblenz on *Monday night, Dec. 10th.* That same night I fell in with a young Englishman at a cigar shop & had a long talk with him. He had been in America, & was related to Cunard of the Steamers.

Next morning, I again rambled about the town — saw the artillery-men & infantry exercise on the parade ground. Very amusing indeed. Saw a squadron of drummers. Walked down & up the river, & while waiting for the Cologne boat, spent at least two hours standing on a stone pier, at the precise junction of the Rhine & Moselle. At 3 o'clock started for Cologne on a Düsseldorf boat. It was intensely cold. Dined at the table d'hôte in the cabin. Fine dinner & wine. Drank Rhenish on the Rhine. Saw Drachenfels & the Seven Mountains, & Rolandseck, & the Isle of Nuns. The old ruins & arch are glorious — but the river Rhine is not the Hudson. In the evening arrived at my old place, Hôtel de Cologne. Recognized Drachenfels in a large painting on the wall. Drank a bottle of Steinberger with the landlord, a Rhinelander & a very gentlemanly well-informed man, learned in wines. At ½ past 6 P.M. went to the Theatre. Three vaudevilles acted. Audience smoking & drinking & looking on. Stopped in a shop on my way home & made some purchases for presents, & was insidiously cheated in the matter of a breastpin, as I found out after getting to London, & not before. God forgive the girl — she was not very pretty either, which makes it the more aggravating.

At 5 A.M. was called & took a chilly breakfast in a vast saloon & paid my bill. The "garçon" who had proved exceedingly polite & attentive to me during my whole [stay] in the house, endeavored in the blandest manner imaginable to cheat me out of one "thaler" in giving me my change. I brought him up however & the villain — the smiling, civil, attentive villain seemed to be sorry that he had forever forfeited my good-will, & in vain!

Drove to the R. R. through an old antique gateway in the city wall. A very cold ride & a very doleful one all the way to Brussels, where we got at 5 P.M., after stopping a while at Aix-la-Chapelle. I saw the old Cathedral & the townhouse — but that was all. At Brussells, went instanter to my hotel there, to get my shirts I had left to be washed. The insidious landlady & the rascally waiters wanted me to stop & dine — & so the pieces "were not come home yet." Had a few comical scenes with the landlady & started off in a pet & went dodging about town to get a cheap dinner. Swallowed a little beer here & ate a bit of cake there, & at last got to the Station house. A horrible long dreary cold ride to Ostend on the coast. In a fit of the nightmare was going to stop at a way-place, taking it for the place of my destination. Arrived at Ostend at 11½ P.M. The boat was at the wharf — took a 2d class passage, & went down into a dog-hole in the bow, & there sat & smoked, & shivered, & pitched about in the roll of the sea from midnight till *5 o'clock in the morning*, when we arrived at Dover.

[67]

Disembarked in the dark in a small boat & went to The Sign of "The Gun." Breakfasted, & took myself off to the Custom House to get my luggage through. They seized my fine copy of "Anastasius" which I had bought at the Palais Royal & told me it was food for fire. Was much enraged thereat. Took 2d class train for London at 6½ A.M. A fat Frenchman in the car with whom I conversed a little. A burly farmer in smallclothes & high boots sat opposite me. Another long dismal ride through Kent, on a raw dismal morning — & at 12 M. arrived at London Bridge — after posting all the way from Coblenz, without cessation. Took bus for Craven Street. On getting there, found that Mr. Rogers had called — "a gentleman from St. James who came in his coach" — as the chambermaid expressed it. Also was handed — with a meaning flourish — a note sealed with a coronet. It was from Mr. Rutland — The Duke of Rutland I mean — inviting me to visit Belvoir Castle at any time after a certain day in January. Can not go — I am homeward-bound, & Malcolm is growing all the time. I was in a villain's garb with travel — had not shaved in a week — dirty shirt &c. So, dressed, & went to Bentley's after letters. Found one from Lizzie, & Allan. Most welcome, but gave me the blues most terribly — Felt like chartering a small boat & starting down the Thames instanter for New York. Dined at the "Blue Posts" & took some punch to cheer me. Came home, had a fire made, & wrote to

Elizabeth Shaw Melville and Malcolm

Lizzie & Allan. While so employed the girl knocked & brought me a package of letters. They were from the Legation, & were from Lizzie & Allan — a week later than those I got in the morning. I read them, & felt raised at once. Both were written in fine spirits — & that was catching. Walked out & read them over again — with the "Powell Papers" at the "Edinburgh Castle." Home early & to bed.

Friday, December 14

Breakfasted at my old place. Then went & bought a Paletot in the Strand, so as to look decent — for I find my green coat plays the devil with my respectability here. Then went & got my hair cut, which was as long as a wild Indian's. Then dressed, & went to Mr. Rogers' — out of town — left card. Thence to Lincoln's Inn Fields & left letter & card for Mr. Foster of the "Examiner." Thence to the College of Surgeons, & called on Mrs. Daniel. Spent a half hour there very pleasantly. Two fine girls, her daughters. (Invited me to tea at 8 o'clock, Saturday). Thence in a bus to the Bank, & stopped to inquire of the agents about the packet ships. I shall go on the "Independence" if nothing happens. Thence home & wrote further to Lizzie & Allan & put up the "Times" for Judge Shaw. Wrote Willis, Duyckinck, & the Judge. Then at 6 P.M. dined at the old place — The Adelphi — & treated myself to a cut of turkey. Thence to the Haymarket & saw Mr. & Mrs. Charles Kean & Buckstone, & Wallack in "The Housekeeper," a drama by

[69]

Douglas Jerrold. The same thing we saw murdered at Canterbury. Left the theater at 9 o'clock & bought a pair of pantaloons for one pound five. Then home & to bed after a cigar — one of the Secretary's bunch.

After breakfast went among the bookstores & stalls about Holywell Street. At last succeeded in getting the much desired copy of Rousseau's "Confessions" for eleven shillings. Walked down Fleet Street after a copy of "Knight's London" — Could not get it — but returning, found a copy in the Strand. Bought it for £1.10. Then home & rigged for Bentley's — whom I expected to meet at 1 P.M. about "White Jacket." Called, but he had not yet arrived from Brighton. Walked about a little & bought a cigar case for Allan in Burlington Arcade. Saw many pretty things for presents — but could not afford to buy. Bought a bread trencher & bread knife near Charing Cross. "The University bread trencher" used of old at Commons, now restored. Very generally used here. A fine thing, & English — Saxon. Home & wrote up journal. At 4 P.M. am to call again at Bentley's. He does not know I am in town — I earnestly hope I shall be able to see him & that I can do something about that "pesky" book.

6 o'clock P.M. Hurrah & three cheers! I have just returned from Mr. Bentley's, & have concluded an arrangement with him that gives me tomorrow his

note for £200. It is to be at 6 months — and I am almost certain I shall be able to get it cashed at once. This takes a load off my heart. The £200 is in anticipation — for the book is not to be published till the 1st of March next. Hence the long time of the note. The above-mentioned sum is for the 1st 1000 copies. Subsequent editions (if any) to be jointly divided between us. I also spoke to him about Lieut. Wise's book, & he is to send for it. At eight tonight I am going to Mrs. Daniel's. What sort of an evening is it going to be? Mr. Bentley invited me to dinner for Tuesday at 6 P.M. This will do for a memorandum of the engagement. I have just read over the Duke of Rutland's note, which I have not fairly perused before. It seems very cordial. I wish the invitation was for next week, instead of being so long ahead — but this I believe is the mode here for these sort of invitations into the country. (Mem. at 1 P.M. on Monday am to call at Mr. Bentley's.)

Sunday, December 16

Last night went in a cab to Lincoln's Inn Fields, & found Mrs. Daniel & "daughts" very cordial. The elder "daught" remarkably sprightly & the mother as nice an old body as any one could desire. Presently there came in several "young gents" of various complexions. We had some coffee, music, dancing, & after an agreeable evening I came away at 11 o'clock, & walking to The Cock near Temple Bar drank a glass of Stout, & home & to bed, after reading a few

[71]

chapters in *Tristram Shandy,* which I have never yet read. This morning breakfasted at 10 at the Hôtel de Sabloniere (very nice cheap little snuggery being closed on Sundays). Had a "sweet omelette" which was delicious. Thence walked to St. Thomas's Church, Charter House, Goswell Street to hear my famed namesake (almost) "The Reverend H. Melvill." I had seen him placarded as to deliver a charity sermon. The church was crowded — the sermon was admirable (granting the Rev. gentleman's premises). Indeed he deserves his reputation. I do not think that I hardly ever heard so good a discourse before — that is from an "orthodox" divine. It is now 3 P.M. I have had a fire made & am smoking a cigar. Would that One I know were here. Would that the Little One too were here. I am in a very painful state of uncertainty. I am all eagerness to get home — I ought to be home — my absence occasions uneasiness in a quarter where I most beseech heaven to grant repose. Yet here I have before me an open prospect to get some curious ideas of a style of life, which in all probability I shall never have again. I should much like to know what the highest English aristocracy really & practically is. And the Duke of Rutland's cordial invitation to visit him at his Castle furnishes me with just the thing I want. If I do not go, I am confident that hereafter I shall upbraid myself for neglecting such an opportunity of procuring "material." And Allan & others will account me a ninny. I would not debate the matter a moment, were it not that at least three whole weeks must elapse ere I start for Belvoir Castle — three weeks! If I could but get over *them!* And if the two images would only *down* for that space of time. — I must light a second cigar & resolve it over again.

1/2 past 6 P.M. My mind is made; rather is irrevocably resolved upon my first determination. A visit into Leicester would be very agreeable — at least very valuable, in one respect, to me — but the Three Weeks are intolerable. Tomorrow I shall go down to London Dock & book myself for a stateroom on board the good ship Independence. I have just returned from a lonely dinner at the Adelphi, where I read the Sunday papers. An article upon the "Sunday School Union" particularly struck me. In an hour's time I must go to Morley's & call upon Stevens & Davis. Would that I could go home in a Steamer — but it would take an extra $100 out of my pocket. Well, it's only 30 days — one month — and I can weather it somehow.

Monday, December 17

Was delighted & exhilarated this morning by a view of the sun — a rare sight here — looking into my river window. Upon sallying out, I found a fine day. Last night after writing my journal (as above) I called on Stevens & sat a while. He spoke of Powell, & that certain persons had called upon him denouncing Powell as a rogue — Poor fellow — poor devil — poor Powell! Took my breakfast as usual at the old place

[73]

& walked down to London Dock. The "Independence" proves to be an old ship — she looks small & smells ancient. Only two or three passengers engaged. I liked Captain Fletcher, however. He inquired whether I was a relation of Gansevoort Melville, & of Herman Melville. I told him I was. I engaged my passage & paid £10 down. Took a bus & rode to Mr. Bentley's according to appointment at 1 o'clock. We concluded the arrangement of "White Jacket" & he gave me his note for £200 at 6 months. From thence went to Mr. Murray's. Had a talk with him about sundry things. On going away, his cousin invited me to dine with him on Wednesday at his chambers in the Temple. Accepted. Thence to the National Gallery & spent an hour looking at Rembrandt's Jew & the Saints of Taddeo Gaddi, & Guido's Murder of the Innocents. Looked in at the Vernon Gallery. Thence walked down the Strand & stopped at a book auction in Wellington Street. Gruff chap had no "catalogues." Thence to Adelphi & dined. Thence to Davidson's — not in. Thence down Newgate St. looking in at the book stores. Saw many books I should like to buy — but can not. Then through Farringdon Street (where I bought pocket Shakespeare, &c.) to the Strand (a little this side of the Bar) & had a long chat with a tobacconist of whom I bought 3/4 pound of very fine Cuba cigars at about 2 cents & a half (our money) a piece. Thence home; & out again, & took a letter for a Duke to the Postoffice & a pair of pants to be altered

[74]

to a tailor. Drank a pint of ale, & by the Haymarket, & so home & wrote Journal.

Miserable rainy day. Treated myself to a sugar omelette at the old place, for breakfast. Thence went to the British Museum — closed. Thence among the old bookstores about Great Queen Street & Lincoln's Inn. Looked over a lot of ancient maps of London. Bought one (A.D. 1766) for 3 & 6 pence. I want to use it in case I serve up the Revolutionary narrative of the beggar. Thence into Chancery Lane, into a horrible hole & bought a fine copy of Chatterton & a 3-Vol. edition of Guzman (Chatterton was 5 Shillings — Guzman, 2). Left them to be bound. Thence to Paternoster Row & saw Davidson. Saw the Literary World's review of "Redburn." Also the Publisher's Advertisements in the New York *Courier*. Gave Davidson my £200 note to get discounted, if possible, at his banker's. I have no receipt to show — so if he dies tonight I am minus $1000. Ditto if he decamps. But he is a very fine, good-hearted fellow — & I hope he won't see this. Upon getting home at 2 P.M. found my copy of "Knight's London" & a note from Mr. Rogers inviting me to breakfast for Thursday next. Accepted. Putting my reply in the office, I stopped at a silversmith's (corner of Craven St. & Strand) & bought a solid spoon for the boy Malcolm — a *fork*, I mean. When he arrives to years of mastication I shall invest him with this fork — as of yore they did a young knight, with his good sword. Spent an hour or so looking over "White Jacket" preparatory to sending it finally to Bentley — who, though he has paid his money, has not received his wares. At 6 I dine with him.

Wednesday, December 19

Last night dined with Mr. Bentley, and had a very pleasant time indeed. I begin to like him much. He seems a very fine frank off-handed old gentleman. We sat down in a fine old room hung round with paintings (dark walls). A party of fourteen or so. There was a Mr. Bell there — connected with Literature in some way or other. At all events an entertaining man and a scholar — but looks as if he loved old Port. Also, Alfred Henry Forrester ("Alfred Crowquill") the comic man. He proved a good fellow — free & easy — & no damned nonsense, as there is about so many of these English. Mr. Bentley has one daughter, a fine woman of 25 & married, and 4 sons — young men. They were all at table. Some time after 11 went home with Crowquill, who invites me to go with him Thursday & see the Pantomime rehearsal at the Surrey Theatre. After breakfast this morning called in at Stibbs' the bookseller, & coming across a fine old copy of Sir T. Browne — bought him, for 16 Shillings (sterling) — about $4. Also, nice edition of Boswell's Johnson, for 21 Shillings. Thence went to the British Museum, & wan-

[76]

dered about for a couple of hours there. Thence to Gordon Square and left letter & card for Mr. Atkinson. Thence through the New Road to Tottenham Court Road into Oxford Street, & so to St. James's. Stopped in Arlington Street to see Mr. Moore the apothecary. Was out of town. Thence to No. 1 St. James's place where I remained about 20 minutes. Thence home & made up a fire at 3 P.M. Tonight at 6 I dine with Mr. Cook.

Thursday, December 20

Last night dined in Elm Court, Temple, and had a glorious time till noon of night. A set of fine fellows indeed. It recalled poor Lamb's "Old Benchers." Cunningham, the author of Murray's London Guide, was there, & was very friendly. A comical Mr. Rainbow also, & a grandson of Woodfall, the printer of Junius, and a brother-in-law of Leslie, the painter. Leslie was prevented from coming. Up in the 5th story we dined. The Paradise of Batchelors. Home & to bed at 12. This morning breakfasted with Mr. Rogers at 10. A remarkable looking old man truly. Superb paintings. No one but he & I together. At leaving he invited me to breakfast with him again on Sunday & meet some ladies. Accepted. Thence went to Bentley's — saw him — got some books out of him. Thence to Mr. Murray's & saw him. With Mr. Cooke I then went to the Erechtheum Club House in St. James Square, where I dine tonight with Cooke's brother, a barrister with a quizzical eye. Thence looked in at the Reform Club House, thence down Whitehall St. & on through the Privy Gardens, & Sir Robert Peel's to the New Houses of Parliament. After much trouble & waiting, got into the House of Lords to see the frescoes. An artist, a friend of Murray's, Mr. John Tenniel, was our "open sesame." Went all over the place — & spent two hours there — much against my will. A finished frescoe by "Herbert" is very fine in the Poets' Hall — representing Cordelia & Lear. Cope is another of the artists. Tenniel's painting is St. Cecilia from Dryden. Thence crossed the river to the Surrey Theater to keep appointment with "Crowquill." Too late. But went in behind the scenes, a little. Thence to Blackfriar's bridge, & took steamer for home. Made a fire & here I am. *Stone*, I believe, is the name of the brother-in-law of Leslie. He was extremely urgent for me to spend Christmas with him & Leslie in St. John's Wood. I could hardly resist him. But I sail on Monday. Last night — just on the eve of my going to the Temple — a letter was left for me — from home! — All well & Barney * more bouncing than ever, thank heaven. In a few days now my letter will be received announcing my sailing. When at Bentley's this morning he said Mr. Miller wanted to see me. Spoke to B. about the time of bringing out *White Jacket*. And we mutually appointed the 23^{d} day of January next (Wednesday) as the day for publishing *here*.

* *Baby boy.*

Last night dined at the Erechtheum Club St. James's with Mr. Cooke (he with the eye) — nine sat down — fine dinner — Rainbow & others were there. Mr. Cleaves is a fine fellow. An exceedingly agreeable company. The "Mall" after dinner, looked at some billiard playing awhile, & left. After breakfast this morning called on Mr. Cleaves at his rooms in the Temple, & we visited the Library — Hall of the Benchers — Kitchen — rooms — Dessert room & *table*. Portraits of the Benchers. In the Library saw some fine old Mss — of the Kings & Queens & Chancellors hundreds of years ago. Thence to Lincoln's Inn, & visited the New Hall, Kitchen, Library &c. — Very fine. Sublime Kitchen — *chimney place*. Also visited the courts in the Inn — 2 Vice Chancellors courts — and the Lord High Chancellor Cottenham — an old fellow nearly asleep on the bench. Thence to the Court of the Master of the Rolls — Rolls' Court — a very handsome man — Lord Something, I forget what — Strange story about him. Thence, left Mr. Cleaves & went to London Dock, through the Wharves to see about the ship's sailing, Saw Captain Fletcher. Sails 3 P.M. tomorrow. Thence walked to Davidson's, Paternoster Row — invited him to dine with me tonight. Thence towards home & bought a large carpet-bag for my traps — price 14 Shillings. Found a note from Mr. Foster of the Examiner inviting me to breakfast Sunday morning. Declined being engaged at Mr. Rogers.

[79]

Dined last night with Mr. Davidson at the Blue Posts. Sat there in very pleasant conversation till 11 o'clock — then home. This morning breakfasted at the old place — then returned to my room & packed up. Cabbed it to London Docks & put my luggage aboard to go round to Portsmouth. Ship sailed at 3 P.M. today. Thence to Davidson's to see about my money. While waiting for him, ran out, & at last got hold of "The Opium Eater" & began it in the office. A wonderful thing, that book. Davidson's cheque I got cashed, & went with the "funds" to Baring Brothers, &c. in Bishopgate St. & got a letter of credit on America for £180. Thence walked home — cold dry day — & made fire at ½ past 3 P.M. Tonight at 6 I dine again at the club in St. James's Square. On this account I declined the invitation from Mrs. Daniel to tea, &c. Cooke sent me a note enclosing an order for me to visit the Reform Club House. He is very obliging.

Last night dined at the Erechtheum Club — a party of eight . . . . Charles Knight, the author of London Illustrated, &c., the Publisher of the Penny Cyclopedia, & concerned in most of the great popular publications of the day — Ford, the Spanish Traveller & Editor of the Guide Book — Leslie, the painter — Cunningham, the London Antiquarian & author of the London Guide published by Mur-

ray . . . . & Mr. Murray the Albemarle Street man —
together with Cooke & a youth whose name I forget.
We had a glorious time & parted at about midnight.
This morning breakfasted with Mr. Rogers again.
And there met "Barry Cornwall," otherwise Mr.
Proctor, & his wife — and Mr. Kinglake (author of
Eöthen?). A very pleasant morning we had, and I
went away at ¼ past one o'clock. Thence walked
through St. James's Park & came home & made a fire.
3 P.M. — While sitting in my room reading the
"Opium Eater" by the fire I am handed Mr. George
Atkinson's card — the girl (pursuant to my directions) having told him at the door that I was "not
in." I am obliged to employ this fashionable shift
of evasion of visitors — for I have not a decent room
to show them — but (& which is *the* cause) I cannot in conscience ask them to labor their way up to
the 4th story of a house. ½ past 3 P.M. Have just
this moment finished the "Opium Eater." A most
wondrous book.

Portsmouth, Monday, December 24, Coffee Room of the Quebec Hotel

After finishing the marvellous book *yesterday*, sal-
lied out for a walk about dusk, & encountered Capt.
Fletcher who was on the point of calling upon me.
Walked down the Strand with him & left him at the
London Coffee House. He seems a very fine fellow.
Dined at Morley's at six P.M. yesterday with Parker
(a little chap) & Somerby, my future fellow passen-
ger in the ship. Home by 11 o'clock & to bed. This

[81]

morning (Monday) breakfasted for the last time at "the old place" and took an affectionate & melancholy adieu of my two ladies there. Thence to Mr. Bentley's — saw him — exchanged my odd volume of the thing — and thence went to the "Reform Club." Superb hall of pillars, &c. Kitchen — pastry room — meat room — cutlets arrayed like cravats — "butters," &c. Bathing & Dressing rooms — boot pullers. Thence strolled about & bought some books — & a bread trencher & knife for Mrs Shaw (£3.10.). Thence to dinner at "the old place" & bid that also an adieu. Bought some things at the Bazaar in the Strand. Bid goodbye to my room in Craven Street — drove to the Waterloo Station in a cab. After a five hours ride, here I am waiting the ship in Portsmouth. Mr. Somerby came with me.

Tuesday, December 25

Christmas.

Rose from a right comfortable English bed and took a short stroll into the town. Passed the famous "North Corner." Saw the "Victory," Nelson's ship, at anchor. While at breakfast with Captain Fletcher, a messenger arrived saying that our ship was off the harbor. Instantly the coffee cups were capsized, & everything got ready for our departure. Jumped into a small boat with the Captain, Somerby, an Englishman (Mr. Jones) and his little girl — & pulled off for the ship about a mile & a half distant. Upon boarding her, we at once set sail with a fair wind, & in less than 24 hours passed the Land's End & the Scilly Isles — & standing boldly out on the ocean, stretched away for New York. — Five days have elapsed — the wind has still continued favorable, & the weather delightful. No events happen — & therefore I shall keep no further diary. I here close it, with my departure from England, and my pointing for home.

At sea, December 29, 1849.

Wednesday, January 30, 1850.

Got sight of a pilot boat, this morning about 12 M.

Captain A. T. Fletcher.
1st mate, Pendleton (?) — 2d. mate, Stokes   *Independence*

Captain Griswold, 1st mate Smith. *Southampton*
2d mate

Memoranda of Things on the Voyage

The pirate & the missionary, —— "all right" with a knowing look, wife on his arm (Capt. Fletcher).

Smuggling grog — mustard bottle on his leg — skin in meat.

(Fletcher)

[83]

Fletcher's story about the catamount pursuing the runaway sailor through the horrible woods of Peru along the seacoast. Leaping from bough to bough. — He slept with his knife in his hand. (Recalled the "Opium Eater" & "crocodile.")

Rousseau a schoolmaster. (2ᵈ Vol, near beginning "could have killed his scholars sometimes."
Dr. Johnson an usher — "intolerable" (vol. 1st.)
(Other allusion in 1st vol. Boswell.)
   3d. Vol. P. 220 — Do.

"A finer man never broke the world's bread" — "If you have friends on the lookout, I have a friend at the helm." — "Did that hamper of wine & those two hams send that man to jail?"

Allusion to *Cannibals* 17th page of Sir Thomas Browne. "Vulgar Errors" (Also 57th page of B. Jonson.)

Indian (Gay-Head) Sweetheart flogged.

Sir Richard Q. C.

A Dandy is a good fellow to scent a room with.

A comical duel Ben Jonson — page 51 ("Out of his humor") Fly — wounding our clothes.

Talk as much folly as you please — so long as you do it without blushing, you may do it with impunity — — (B. J. p. 54).

	£	S.	D
√ Ben Jonson. folio 1692		.13.	
Davenant. " 1673		.10.	
√ Beaumont & Fletcher folio		.14.	
Hudibras 18 mo. (old)		2.	
√ Boswell's Johnson (10 vol. 18 mo.)		21.	
√ Sir Thomas Browne folio. 1686			
3 copies of Mardi, from Bentley			
2 copies of "Redburn" from Bentley			
Guide Book for France	} From Mr. Murray		
" " " Germany			
Knight's London (3 vol oct.)	1.	10.	0
Lavater		10.	
√ Rousseau Confessions		11.	
Castle of Otranto		1.	
2 plays of Shakespeare		.	2
√ Charles Lamb's works (octavo)	} From Mr. Moxon		
√ Final Memorial of Lamb			
√ Guzman 3 vol.		3.	
√ Chatterton 2 "			
Old Map of London (1766.)		3.	6
√ Anastasius (2 vol.) Bentley			
√ Caleb Williams (1 vol.) Do			
Vathek (1 vol.) Do			
√ Corinne " Do.			
√ Frankenstein " Do.			
Aristocracy of England Bow Street	} 5.		
Marlowe's Plays ' Do.			

[85]

Autobiography of Goethe   (Bohn)   3.
Letters from Italy (Goethe)   Do.   3.
Confessions of an Opium Eater   1   6.

Books obtained in Paris

Telemachus   —   About 2 francs
Anastasius (2 Vol.)   "   4   "
Views of Paris.   (R. R. Station.)

Books obtained in Germany

"Lays & Legends of the Rhine"   Coblenz.
"Up the Rhine"   Cologne
Panorama of the Rhine   "

Paris

"Curios"

Medal of Napoleon & Josephine   5 francs
   "   (Battle)   1   "

Porcelain stoppers   —Cologne.
         2½ Groschen (6 pence)

Monday 19 Nov.
    6½ P.M.   Haymarket with Mr. Langford.

Tuesday 20th

Mrs. Lawrence, Clarendon, 8 P.M. (call)

Wednesday 21

9½ P.M.   Langford   3 Furnival Inn
        Supper.

Books obtained in London 1849

1692

	£	s	D
✓ Ben Jonson. folio		.13.	
Davenant , "		.10.	
✓ Beaumont & Fletcher folio		.14.	
Hudibras 18 mo. (old)		2.	
✓ Boswell's Johnson (10 vol. 18 mo)		21.	
✓ Sir Thomas Browne folio 1696			

3 copies of Mardi, from Bentley
2 copies of "Redburn" from Bentley
Guide Book for France } from Mr.
" " " Germany } Murray.

Knight's London (3 vol oct:)	1.	10.	0
Lavater		10.	
✓ Rosseau Confessions ('		11.	
Castle of Otranto		1.	
2 plays of Shakspeare			.2
✓ Charles Lamb's works (octavo) } from Mr.			
✓ Final Memorials of Lamb } Moxon.			
✓ Gusman 3 vol.		3.	
✓ Chatterton 2 "		3.	6
Old Map of London (1766.)			
✓ Anastasius (2 vol) Bentley			
✓ Caleb Williams (1 vol) Do			
Vathek (vol) Do			
✓ Corinne " Do			
✓ Frankenstein " Do			
Aristocracy of England Bon Street }		5.	
Marlowe's Plays Do)			
Autobiography of Goethe (Bohn)		3.	
Letters from Italy (Goethe Do.		3.	
Confessions of an Opium Eater.	1	6.	

Facsimile — see page 85

Thursday 22

Friday 23

Murray's   ¼ before 7 — dine, Lockhart.

Cannibals
Execution
Terrapin

M. F. Tupper — Albany, Guildford
By London Bridge — about 20 miles (Hatchards')
Lord John Manners — town address Nº 3 Albany. —
Belvoir Castle, Leicester.
Longmans   37 Paternoster Row.
Mr. Langford Furnival's Inn. (3)
Mulligan (Wᵐ) 45 Walker St. — N.Y.
Lady Elizabeth Drummond 2 Bryanstone Square,
& Cadland, Southampton.
U. S. Legation 1 Upper Belgrave Street.
J. Bates 46 Portland Place.
   Near Park Crescent.   New Road
Russell Sturgis 27 Upper Harley   omnibus
   Street
Edward Moxon 44 Dover Street, Piccadilly. Nearly
   opposite St. James's Street — back of Murray's.
G. Adler No. 3 Rue de la Convention, Paris.
Albert Smith 14 Percy Street, Bedford Square.
J. Bates (country address) East Sheen, Surrey.

[87]

Duke of Rutland's is in Leicester.

R. M. Milnes Fryston, Ferrybridge, Yorkshire.

Richard Bentley 29 Broad Street, Brighton.

Robert Francis Cooke. 4 Elm Court, Temple.

Alfred Henry Forrester ("Crowquill") 3 Portland Place, North Clapham Road.

John Miller, Tavistock Street, Covent Garden.

Mr. Cleaves — 9 King's Bench Walk, Temple. (11 o'clock, Friday).

John Foster 58 Lincoln's Inn Fields.

George C. Rankin Esq. 25th Reg Nat. Inft. Hadjypoor, Punjab.

14 *s.* for spoon

L. J. Manners Belvoir Castle, Leicester

NOTES

*However, there being several particular Persons
— which are not commonly known, and some
old Stories — which want explication, we have
thought fit to do that right to their Memories,
and for the better information of the less learned
Readers, to explain them in some additional
Annotations —* Zachary Grey, Preface to Samuel
Butler's *Hudibras* (1764).

ABBREVIATIONS

HCL–M	Harvard College Library — Melville Collection
NYPL–D	New York Public Library — Duyckinck Collection
NYPL–GL	New York Public Library — Gansevoort-Lansing Collection
NYPL–B	New York Public Library — Henry W. and Albert A. Berg Collection
MHS–S	Massachusetts Historical Society — Shaw Papers
MHS–D	Massachusetts Historical Society — Dana Papers
MHS–E	Massachusetts Historical Society — Everett Papers
DNB	Dictionary of National Biography

NOTES

The words "Herman Melville's" were added at the head of the Journal by Melville's wife, Elizabeth Shaw Melville.

Thursday, October 11

West. "Melville left on Wednesday week. The morning was so wet and rainy that none of his friends went to the Hook with him, but the wind was good and must have carried him on well." — Letter from George Duyckinck to Miss Joan Miller, October 18, 1849.

He had previously written her, October 5, 1849:

Herman Melville sails on Monday for London. We shall miss his society here but I have no doubt his journey will be of great service to him. . . . — NYPL–D

The "Southampton." The *Southampton* was a packet boat, a three-deck square-rigged sailing liner of 1299 tons which made the regular run from New York to London. These ships carried cargo, mail, and passengers, although by the time Melville made this journey, the faster steamers were already taking the ascendency. The old-fashioned packet, however, was still fairly popular: the captain lived in close society with the passengers, and they all shared the vicissitudes that frolic winds might cause upon the sails. A farmyard including cows, sheep, pigs, ducks, and hens lived upon the hatches to keep the table freshly supplied with milk, eggs, and poultry.

Naval architecture has provided few models as beautiful as the full-rigged ship, with its three masts and billowing sails. The packet partook of much of the grace of the regular clipper ship, except that its hull bellied out further on each side in order to make more room for the cargo.

I am indebted to Ruth Whitman for this note.

The "Narrows."

At last we got as far as the Narrows, which everybody knows is the entrance to New York Harbour from sea; and it may well be called the Narrows, for when you go in or out, it seems like going in or out of a doorway; and when you go out of these Narrows on a long voyage like this of mine, it seems like going out into the broad highway where not a soul is to be seen. — *Redburn.*

Allan. Allan Melville (1823–1872), his younger brother, a lawyer with offices in Wall Street.

George Duyckinck. George Long Duyckinck (1823–1863), younger brother of Evert Augustus Duyckinck, and coeditor with him of the *Literary World, a Journal of American and Foreign Literature, Science, and Art,* October 7, 1848 to December 31, 1853, and of the *Cyclopaedia of American Literature,* 1855.

The Captain. Captain Robert Harper Griswold (1806–1882). "He was a favorite commander of packet-ships of the John Griswold Line, sailing between New York and London, a man of much reading, and in his prime, of elegant manners and great personal beauty." — *Family Histories and Genealogies,* second volume, containing a series of genealogical and biographical monographs on the families of Griswold, Wolcott, Pitkin, Ogden, Johnson, and Diodati, by Edward Elbridge Salisbury and Evelyn McCurdy Salisbury (New Haven, 1892).

Friday, October 12

Mr. Adler. George J. Adler (1821–1868), philologist, born in Leipzig, brought to the United States in 1833. He is best known as the author of a *Dictionary of the German and English Languages* (1849), the labor of which so impaired his health that he suffered intermittently from mental illness the rest of his life. On this trip he was going abroad for rest and relaxation.

Mr. Taylor. Dr. Franklin Taylor, who had accompanied his cousin, James Bayard Taylor, on the pedestrian tour which resulted in the latter's popular *Views A-Foot,* as noted in an unpublished dissertation on *Clarel* at Yale by Walter H. Bezanson (1943).

James Bayard Taylor (1815–1878). "His age's young hero among travellers." He wrote the valentine for Melville at the party given by Miss Anne Lynch, February 14, 1848, which was published in N. P. Willis's *Home Journal,* March 4, 1848, under the heading *The Valentine Party.*

> *Bright painter of those tropic isles,*
> *That stud the blue waves, far apart,*
> *Be thine, through life, the summer's smiles,*
> *And fadeless foliage of the heart:*
> *And may some guardian genius still*
> Taboo *thy path from every ill.*

I am indebted to Mr. Luther S. Mansfield for this transcription of the valentine, and information of its publication. It is also reproduced in his unpublished dissertation, Chicago, *Herman Melville, New Yorker.*

Saturday, October 13

Fixed Fate, Free will, foreknowledge absolute. See Melville's pattern of the problem in the chapter entitled "The Mat Maker," *Moby Dick.*

close-reefed. Originally *double-reefed.*

dim stars in the sky. See the chapter entitled "The Candles," *Moby Dick,* where the corposants are "like a far-away constellation of stars."

Sunday, October 14

I alone am left. The language of the King James *Job* had evidently already made its mark, and Melville could make a serio-

[93]

comic use of it here. See the quotation at the head of the Epilogue in *Moby Dick*.

Mrs. Kirkland. Mrs. Caroline M. S. Kirkland (1801–1864) author of *Holidays Abroad or Europe from the West*, Baker and Scribner (New York, 1849); but perhaps better known for her early writings on pioneer life in Michigan. According to Poe, Mrs. Kirkland "was frank, cordial, yet sufficiently dignified, — even bold, yet especially ladylike; converses with remarkable accuracy as well as fluency; is brilliantly witty, and now and then not a little sarcastic, but a general amiability prevails."

Mrs. Kirkland was one of the guests invited to Miss Anne Lynch's Valentine party, February 14, 1848, for which Bayard Taylor wrote the valentine for Melville. He probably met her then.

Omoo. The sequel to *Typee*, published 1847.

Dr. Armsby. James H. Armsby, M.D., professor of Anatomy and Physiology, Medical College, res. at 669 Broadway. — *Albany City Directory 1849*.

Mr. Twitchell. Asa Weston Twitchell (1820–1860) lived in Troy and Albany. In 1848 he painted a portrait of Dr. Armsby's son. — *National Academy of Design Exhibition Record 1820–1860*. Printed for the New York Historical Society.

The portrait of Melville reproduced here for the first time has been identified by Mr. J. D. Hatch, Jr., Director of the Albany Institute of History and Art, as the work of Twitchell, and is undoubtedly the portrait referred to. It can be dated at some time between 1845 and 1847, during Melville's residence with his mother in Lansingburg on his return from the South Seas.

Monday, October 15

Eastern jaunt. Early in September Melville had visited his friend, Evert Duyckinck, and proposed that he join him in "a flying trip of eight months, compassing Rome"; and on October 6 he had written Richard H. Dana, Jr. that *White Jacket* would

[94]

not be out for several months, and that he expected to be away several months after its publication. Now he was hopefully making plans for extending his itinerary.

But this trip was not taken till 1856–57, and then alone, and with spirits not so buoyant. See *Journal Up The Straits*, edited by Raymond Weaver, published by the Colophon (New York, 1935).

"Powell." Presumably the same "Powell" mentioned twice again in the *Journal,* December 13 and December 17. Thomas Powell (1809–1887), poet, dramatist, journalist, who came to America in the Spring of 1849, "a bluff, hearty, lively Englishman, fond of genial companionship, good beer and conversation." He became one of the "Knights of the Round Table" that met at the Duyckincks', 20 Clinton Place, and a member of the coterie of New York journalists and literary men that assembled regularly at Pfaff's, 653 Broadway, where he was "one of the liveliest and best-liked." Among his works are a play, *The Blind Wife,* 1843; *Living Authors of England,* 1849; and *Living Authors of America,* 1850.

Tuesday, October 16

Little Barney. In Melville's hand is a list of "What became of the ship's company of the whale-ship *Acushnet,* according to Hubbard who came home in her (more than a four years' voyage) and visited me at Pittsfield, in 1850." Barney appears there as a boat-steerer. This it seems likely was a nickname for Wilson Barnard, whose name appears on the official "List of Persons Composing the Crew of the Ship Acushnet of Fairhaven where-of Master Valentine Pease — bound for Pacific Ocean." He is described as being five feet and a half an inch tall — certainly a "little" Barney. Lacking any other more convincing source of Melville's nickname for his son, this one may fairly be accepted. — Melville's list of 1850 (HCL–M). The official crew list is at the Old Dartmouth Historical Society and Whaling Museum, New Bedford.

Orianna. Orianna (or Oriana) was one of Melville's nicknames for his wife. Oriana, the mistress, and later the wife, of Amadis de Gaul, was noted for her loyalty. Amadis at different periods of his life called himself "The Child of the Sea," "Beltenebros" (the fair forlorn), "The Knight of the Green Sword," and "The Greek Knight." See *Amadis of Gaul,* attributed to Vasco Lobiera, late thirteenth century. Tennyson's *The Ballad of Oriana* had appeared in many editions of his poems from 1842 to 1849. It had just been reprinted in Thomas Powell's *Living Authors of England* published in 1849 by D. Appleton & Company, New York, and George S. Appleton, Philadelphia.

If Melville was not familiar with the original thirteenth century legend (and it is not known that he had not read it) it is fair to assume he knew Tennyson's ballad. His sister, Augusta, owned the 1842 edition of *Tennyson's Poems* in two volumes, published by William D. Ticknor, Boston, the second volume of which is inscribed with her name, "From Mrs. Blatchford, Dec: 19ᵗʰ 1844." These volumes, now in my possession, were given to her brother Thomas after Augusta's death in 1876. One verse of the ballad may be interesting to the reader:

> *My heart is wasted with my woe,*
> > *Oriana.*
> *There is no rest for me below,*
> > *Oriana.*
> *When the long dun wolds are ribb'd with snow,*
> *And loud the Norland whirlwinds blow,*
> > *Oriana,*
> *Alone I wander to and fro,*
> > *Oriana.*

Saturday, October 20

Mrs. Gould. Mrs. N. W. Gould on the passenger list. Undoubtedly Mrs. Julia Gould who made her debut at the Lyceum Theatre, London, in 1842, and her first public appearance in

America in September, 1850, in New York. These notices of her appearances are from the *History of the American Stage* by T. Allston Brown (Dick and Fitzgerald: N. Y., 1870).

Sunday, October 21

The Banks. A shelving elevation in the sea off the coast of Newfoundland, known for its excellent fishing, but notorious for foggy, stormy weather, "Newfoundland weather."

Thursday, November 1

Lizzard. The southernmost point of Cornwall.

Friday, November 2

White cliffs indeed. One foreshadowing, perhaps, of the association of whiteness with "things the most appalling to mankind." See the chapter on "The Whiteness of the Whale." *Moby Dick.*

Saturday, November 3

The Needles. A cluster of three pointed rocks in the English Channel, West of the Isle of Wight. — *Lippincott's Pronouncing Gazetteer of the World.*

Sunday, November 4

in tumbler. The usual nautical expression is "in gimbels." "Tumbler" may have been a term used as Melville used it at the time, to designate the mechanical device which kept a lamp in a vertical position in spite of the motion of the ship. It has been brought to my attention by Mr. Whitehill of the Boston Athenaeum, an authority on nautical matters, that many words in common or popular use never reach a dictionary. This word did have *other* nautical definitions, however, found in dictionaries, with which Melville must have been familiar. He may also have had stored in his memory the experience of another kind of tumbler "a toy, usually representing a grotesque figure, having

[97]

the centre of gravity low and the base rounded so as to continue rocking when touched" (and remain vertical when at rest).

lapse of ten years. Melville made his first trip to England in 1839, as a sailor aboard a Liverpool packet. He utilized material from this trip in *Redburn.*

Peedee. A river (more correctly, the Pee Dee) near Georgetown, South Carolina. It is probable Melville was familiar with it through Richard Lathers, whose early home was in Georgetown, and who had married a sister-in-law of Melville's brother Allan. Peedee was his satirical substitution for *Typee;* Hullabaloo, for *Omoo;* Pog-Dog for *Redburn.*

stout (Dinner). These memoranda are jotted on the inside of the back cover of the "small paper book" in which Melville started his journal.

Monday, November 5

Melville several times made mistakes in the date or day of the week. He had actually marked this day Monday, November 6, despite the fact that he dated the purchase of the *Journal* Wednesday, November 7. Tuesday is dated 6 *and* 7, one superimposed on the other. As explained in the Preface, these have been corrected for the sake of consistency.

Cinque Ports. Sandwich, Dover, Hythe, Romney, and Hastings were the original five members of an association organized on French lines in the thirteenth century to furnish the ships and men of the king's navy. Deal was only a "limb" of the original five according to *Murray's Handbook for Kent* (1877).

An imposing ruin. The walls of the fort "some 11 ft. thick, which remain to an impressive height, as much as 25 ft. in places, are without doubt one of the finest examples of Roman masonry still existing in this country." — *The Journal of the British Archeological Association,* v. IV (1849).

circus. A Roman circus, being uncovered in 1849 by the antiquary, Mr. William Henry Rolphe of Sandwich. "An amphitheatre, 200 ft. by 116 ft., situated on the southern side, was excavated in 1849, . . . and probably belongs to an early period, as a burial with a coin of Constans 332–50 A.D. was discovered over the remains of the west entrance, suggesting that the structure had fallen into decay long before that date." — *The Journal of the British Archeological Association,* v. IV (1849).

the proprietor. Mr. William Henry Rolphe of Sandwich

. . . . is to Richborough the tutelary *genius loci.* The zealous labours of such persons are but little appreciated by the world, and their example is seldom followed; and those who are best adapted by circumstances to observe and collect facts which illustrate ancient local history, are usually the least qualified, by taste or by education, to understand and apply them; the wants and enjoyments of the present, and the worldly hopes of the future, constitute the business of life, and the past is regarded only as a dry and obsolete book, whose pages of wisdom, to the busy actors in the present scene, are dull and unprofitable. — *The Antiquities of Richborough, Reculver, and Lymne, in Kent,* by Charles Roach Smith, F.S.A., illustrated by F. N. Fairholt, F.S.A.

The dedication of this book, signed by both the author and the illustrator, gives an idea of the kind of man Melville talked with.

To
William Henry Rolphe, Esq.
of Sandwich.
as a tribute of esteem for his zeal in investigating and preserving the antiquities of his neighborhood and native country,
as well as for the liberality with which he affords access to his collections, and encourages the researches of others,
this volume is inscribed,
with the best wishes of his sincere friends,
Charles Roach Smith,
Frederick W. Fairholt.

London
July 1, 1850.

Mr. Rolphe had made rich finds of pottery, glass, personal ornaments, wall paintings, implements and utensils, animal remains, coins, and miscellaneous articles, which he had collected in his museum in Sandwich.

That Melville was nourished by the past rather than scornful or negligent of it, is almost too obvious to mention to anyone familiar with his writings. It is also obvious in the books, prints, and *objets d'art* with which he surrounded himself. But that the antiquarian spirit *per se* was not his, he had already demonstrated in two satirical chapters in *Mardi*, "They Visit An Extraordinary Old Antiquary," and "They Go Down Into The Catacombs."

Dane John. A lofty mound close within the city walls, that may have had some connexion with the castle beyond; or may mark the sight of some earlier British stronghold. The name is no doubt a corruption of *Donjon. — Murray's Handbook for Kent* (1877).

theater. Melville struggled with the spelling of this word, starting with *theater,* acquiring *theatre,* and ending with *theatere* for certainty and full measure. Names of particular theatres he sometimes spelled correctly, sometimes in his native fashion. I have given them their accustomed form, but have let his freedom of spelling stand in the use of the word elsewhere, except for the superfluity of e's in the two instances where they occur.

Incidentally, the play he saw that night was *The Housekeeper* by Douglas Jerrold.

Tuesday, November 6

Penshurst. The home of Sir Philip Sidney.

> . . . *thou wilt tread*
> *As with a pilgrim's reverential thoughts*
> *The groves of Penshurst. Sidney here was born.*
> *Sidney, than whom no greater, braver man*
> *His own delightful genius ever feigned*
> *Illustrating the groves of Arcady*
> *With courteous courage and with loyal love.*
> — Robert Southey, *For A Tablet At Penshurst.*

Tunbridge (fine old ruin there). The Castle, of the Early Decorated period, 1280–1300, with fragments of Norman and Early English work. — *Murray's Handbook for Kent and Sussex* (1858).

Concerts. Popular concerts, inaugurated and led by Louis Antoine Julien (1812–1860), Swiss-born, eccentric, lavish conductor, and composer of popular dance music who did much to educate the taste of his vast audiences by fine performances of the best symphonies and overtures. The program Melville heard was long and varied: it included such different numbers as Mendelssohn's *Grand Symphony in A Minor, Home Sweet Home,* and the *Cossack Polka founded on the Siberian Melodies*. In a somewhat patronizing tone, the *London Times,* December 13, 1849, offers congratulations and encouragement:

> M. Julien may be congratulated on having made considerable progress lately in the right direction. His Beethoven and Mendelssohn nights have been eminently successful, and may be said to have created a new and wholesome appetite in the minds of that part of the public on whose patronage he mainly depends.

Redburn. This was a favorable review of a book of which Melville himself thought slightingly. It begins:

> Indebted less for its interest to the regions of the fantastical and the ideal, than to the more intelligible domain of the actual and real, we are disposed to place a higher value upon this work than upon any of Mr. Melville's former productions. Perhaps it is that we understand it better, and the fault is not in Mr. Melville, but in ourselves, that we appreciate more satisfactorily the merits of a story of living experience than the dreams of fancy and the excursions of a vivid imagination. There are occasional snatches even in this story of the same wild and visionary spirit which attracted so much curiosity in its predecessors, and they come in with excellent effect to relieve and heighten its literal delineations; but the general character is that of a narrative of palpable life, related with broad simplicity, and depending for its final influence over the sympathy of the reader upon closeness and truthfulness of portraiture. In the Dutch fidelity and accumulation of the incidents, it is a sort of Robinson Crusoe on shipboard.

[101]

It ends as follows:

The work displays an intimate acquaintance with the mysteries of seamanship, and a rich graphic power in the use and treatment of them. The idiomatic peculiarities of the style, which will enable the reader at once to trace the native source of the authorship, impart a congenial flavour to the whole, which greatly increases that sense of reality which constitutes the paramount merit of the work.

— *Bentley's Miscellany,* v. 26 (1849), pp. 528–530.

The breadth and vision of *Bentley's Miscellany* could not be better shown than in contrasting this review of *Redburn* with their previous review of *Mardi,* which denominates Melville as "a man intoxicated with imagination." It continues:

. . . For feeling in its ordinary shapes he has no toleration, and he thinks, not altogether perhaps without reason, that the world also is growing weary of it. He endeavours, therefore, to imitate one of the most striking processes of civilization, and to build up for fancy a distant home in the ocean. In the development of this design he is guilty of great extravagance; but while floating between heaven and earth creating archipelagoes in the clouds, and peopling them with races stranger and more fantastical than

> — *The cannibals that each other eat;*
> *The Anthropophagi, and men whose heads*
> *Do grow beneath their shoulders,*

he contrives to inspire us with an interest in his creations, to excite our passions, to astonish us with the wild grandeur of his landscapes, and to excite in us a strong desire to dream on with him indefinitely.

The whole review, an interesting piece of writing in itself, ends with:

We recommend the reader to try his luck with "Mardi," and to see whether a trip into the Pacific may not prove quite as agreeable as a lounge through Belgravia. The chances, we think, are in favour of the ocean.

— *Bentley's Miscellany,* v. 25 (1849), pp. 439–442.

Blackwood's. The spirit of this review exhibits none of the humility of that in *Bentley's Miscellany,* but rather the superior attitude that American authors were beginning to resent in their British critics. A scene between Redburn and Captain Riga was found to be "quietly humorous, and reminds us a good deal of

Marryat, in whose style of novel we think Mr. Melville would succeed." — *Blackwood's,* v. 66 (1849), p. 569.

Madame Vestris. Lucia Elizabeth Mathews, (1797–1856) wife of Charles Mathews, had a long career of activity and popularity, first in Italian opera, then in English drama. "Her chief gifts were archness, fascination, *mutinerie,* a careless acceptance of homage, and a kind of confidential appeal to an audience by which she was always spoiled." — DNB

Charles Mathews. Charles James Mathews (1803–1878), actor, had even longer activity and popularity than his wife. He was

"utterly powerless in the manifestation of all the powerful emotions. . . . He cannot even laugh with animal heartiness. He sparkles; he never explodes." Mathews had, however, airiness, finesse, aplomb, and in spite of an occasional tendency to jauntiness, repose and good breeding, which are rare on the English stage, and he had powers of observation and gifts of mimicry.

— DNB

Royal Lyceum Theatre. Under the management of Madame Vestris, there were advertised for that night, *Not A Bad Judge* and *Beauty And The Beast,* a two-act play, and a two-act fairy extravaganza, both by J. R. Planché.

Gallery. In 1853 Melville wrote *The Two Temples,* refused in 1854 by *Putnam's Monthly* (for fear of "offending the religious sensibilities of the public") and first published in 1924 in *Billy Budd and Other Prose Pieces.* In it he used his shilling gallery experience at the Royal Lyceum on November 7, combining it with an imaginary performance of Macready in the part of Cardinal Richelieu.

I stood within the topmost gallery of the temple. . . . This time I had company. Not of the first circles, and certainly not of the dress-circle; but most acceptable, right welcome, cheery company, to otherwise uncompanioned me. Quiet, well-pleased working men and their glad wives and sisters, with here and there an aproned urchin, with all-absorbed, bright face, vermillioned by the excitement and the heated air, hovering like a painted cherub over the vast firmament below.

[103]

. . . I saw a sort of coffee-pot and pewter mug hospitably presented to me by a ragged, but good-natured-looking boy. . . . Out from the tilted coffee-pot-looking can came a coffee-coloured stream, and a small mug of humming ale was in my hand.

— *Billy Budd and Other Prose Pieces.* Manuscript of *The Two Temples.* (HCL — M.)

Thursday, November 8

Mr. Bentley's. Richard Bentley (1794–1871), bookseller and publisher, 8 New Burlington Street, the publisher of *Bentley's Miscellany, Oliver Twist,* and *Barnaby Rudge;* also 127 volumes of "Standard Novels."

Princess's Theatre. The program, as advertised, was Donizetti's opera buffa, *Don Pasquale,* the new extravaganza *The First Night,* and a new ballet, *Les Patineurs. The First Night* is a farce in one act by Tom Parry.

Friday, November 9

Mulligan. Presumably the William Mulligan of 45 Walker Street, New York, listed at the end of the *Journal;* one of the masters of Mulligan and Roberts School, of that address.

one of the Bridges. "In *Israel Potter* (1855) Melville devoted one chapter to a description of London Bridge: a chapter entitled: 'In the City of Dis' . . . [He] had been husbanding for six years the impressions gathered on November 9, 1849." — *Herman Melville, Mariner and Mystic,* Raymond Weaver.

Theatre Royal, Adelphi. The program as advertised was *The Sons of Mars, Domestic Economy, A Bird of Passage,* and *Mrs. Bunbury's Spoons,* all new productions. *The Sons of Mars* was first played on October 22 at the Adelphi; *Domestic Economy,* a farce in one act by Mark Lemon, opened November 8; *A Bird of Passage,* a farce in one act, opened at the Haymarket on September 10; and *Mrs. Bunbury's Spoons,* a farce by Joseph Stirling Coyne, opened at the Adelphi October 15.

Lord Ellenborough. Edward Law, first Baron Ellenborough (1750–1818), lord chief justice of England, was a scholar at the Charterhouse from his twelfth to his seventeenth year.

Duke of Norfolk. Thomas Howard, fourth Duke, executed June 2, 1572 for conspiring with Mary Queen of Scots, the Bishop of Ross, Ridolphi (a Florentine banker), King Philip of Spain, and the Duke of Alva to put Mary on the English throne.

Whittington. Sir Richard Whittington (circa 1358–1423), the celebrated Lord Mayor of London and benefactor of his city. The house Melville was shown was the remains of his palace, not his birthplace, as the friendly member of the Fire Department believed.

"In Hart Street, four doors from Mark Lane, up a gateway, are the remains of the residence of the celebrated Whittington." — Lambert's *History of London* (1806). Quoted in *The Model Merchant of the Middle Ages, Exemplified in the Story of Whittington And His Cat,* Rev. Samuel Lysons, M.A. (1860).

Melville would have relished the medieval address of "Hart Street, Crutched Friars."

(*A good thing might be made of this*). Melville recalled this scene, and made use of it five years later when he wrote *Poor Man's Pudding And Rich Man's Crumbs*:

A few moments more and I stood bewildered among the beggars in the famous Guildhall.

Where I stood — where the thronged rabble stood, less than twelve hours before sat His Imperial Majesty, Alexander of Russia; His Royal Majesty, Frederick William, King of Prussia; His Royal Highness, George, Prince Regent of England; His world-renowned Grace, the Duke of Wellington; with a mob of magnificoes made up of conquering field-marshals, earls, counts, and innumerable other nobles of mark. . . .

One after another the beggars held up their dirty blue tickets, and were served with the plundered wreck of a pheasant, or the rim of a pasty — like the detached crown of an old hat — the solids and meats stolen out.

"What a generous, noble, magnanimous charity this is!" whispered my guide.

[105]

Poor Man's Pudding And Rich Man's Crumbs was first published in *Harper's New Monthly Magazine* (June, 1854), and again in *Billy Budd and Other Prose Pieces* (1924).

Kilmarnoch. William Boyd, fourth Earl of Kilmarnoch, commander of a troop of horse under Charles, the Young Pretender, was beheaded August 1, 1746.

Tunnel. In the tunnel connecting Wapping and Rotherhithe Melville records that he bought

> two medals (warranted not silver) which I wish little Evert & George * to keep by way of remembrances that I remembered them, even while thirty feet under water.

— Letter to Evert Duyckinck, given in full by Willard Thorp in his *Herman Melville*. The original, dated [New York] Saturday Evening, Feb. 2d. [1850] — NYPL–D

> "Hold; while Prometheus is about it, I'll order a complete man after a desirable pattern. Imprimis, fifty feet high in his socks; then, chest modelled after the Thames tunnel" — *Moby Dick*.

Penny Theatre. I am indebted for my information on the Penny Theatres to a bound collection of clippings entitled *Penny Theatres Illustrated with Views, Bills, Advertisements,* etc., by Frederick Burgess (1882), in Harvard's Theater Collection.

> Penny Theatres, or "Gaffs" as they are usually called by their frequenters, are places of juvenile resort in the metropolis which are known only by name to the great mass of the population. . . . Though the number of Penny Theatres in London cannot be counted, it is beyond all question that they are very numerous. They are to be found in all the poor and populous districts.

These theatres were unlicensed, and were variously described by the more fortunate members of the community as "infamous nuisances, loathsome receptacles and holes of iniquity." They gave several performances in an evening, a penny each, which thousands of young people from six to eighteen years old attended each night. These children were "not only fond of extremes, but would tolerate nothing else. The deepest tragedy or the broadest

* Erratum: for "George" read "Henry."

farce were their fare: the more sordid the details, the better." It is not strange that these places were "a hot-bed for breeding young criminals," where "the greatest villain was the greatest hero with the audience."

> The dramatis personae . . . keep up in most cases a very close intimacy with the audience. In many instances they carry on a sort of conversation with them during the representations of the different pieces.

As for the auditorium —

> It is all gallery together. And such galleries! The seats consist of rough and unsightly forms. There is nothing below the feet of the audience; so that any jostling or incautious movement may precipitate them to the bottom. The ascent to the galleries is usually by a clumsy sort of ladder, of so very dangerous a construction, that he who mounts it and descends without breaking his neck, has abundant cause for gratitude!

All this was usually lighted by half a dozen candles.

With such a physical structure, with an audience "reeking with unwholesome damp," with the possibility of fisticuffs between members of the audience and characters on the stage, or between different members of the audience, or even between members of the cast, the apprehensions of a timid soul would not be likely to lessen!

As late as 1882 efforts were still being made to abolish these pitiful sources of entertainment for London's very poor. But they satisfied a legitimate need, wretched and warping as they were.

Melville was not one to be afraid either of wretchedness or ladders.

Sunday, November 11

Crusaders.

> . . . these figures . . . have a majesty which fills the imagination. The faces depicted in the death-calm, are dignified, as death always is; and their character is so various, that one studies them in full confidence that they are portraits of the mighty men of war of the twelfth century. — *Holidays Abroad* by Mrs. Caroline M. S. Kirkland.

Melville read Mrs. Kirkland's account of European travel on board ship (see October 14 of the *Journal*). He may have met her in New York; for, according to Luther Mansfield's unpublished dissertation, *Herman Melville, New Yorker*, she was one of the number who frequented the salon of Miss Anne C. Lynch, later Mrs. Vinzo Botta, many of whom he knew. Melville wrote in 1854:

> Go view the wondrous tombs in the Temple Church; see there the rigidly-haughty forms stretched out, with crossed arms upon their stilly hearts, in everlasting and undreaming rest. Like the years before the flood, the bold Knights-Templars are no more. Nevertheless, the name remains, and the nominal society, and the ancient grounds, and some of the ancient edifices. But the iron heel is changed to a boot of patent leather; the long two-handed sword to a one-handed quill; the monk-giver of gratuitous counsel now counsels for a fee; the defender of the sarcophagus (if in good practice with his weapon) now has more than one case to defend; the vowed opener and clearer of all highways leading to the Holy Sepulchre, now has it in particular charge to check, to clog, to hinder, and embarrass all the courts and avenues of law; the knight-combatant of the Saracen, breasting spear-points at Acre, now fights law-points in Westminster Hall. The helmet is a wig. Struck by Time's enchanter's wand, the Templar is to-day a Lawyer.
> — *The Paradise of Bachelors and the Tartarus of Maids* in *Billy Budd and Other Prose Pieces.*

fine. Both from Mrs. Kirkland's and Melville's own description of these figures, I take it he omitted the word "knights" or "crusaders" unintentionally. He meant to convey the impression that he found their heads "damned fine." The original reads, "Heads of the damned, fine." The comma should obviously have come after the omitted noun.

Chambers. Sir William Chambers (1726–1796) was an architect of eclectic tastes, designer of the main part of Somerset House.

Thomson. James Thomson (1700–1798), author of *The Seasons.*

The Beauties. The most famous work of Sir Peter Lely (1618–1680), a collection of the portraits of the ladies of the Court of Charles II. "The eyes of the ladies are drowsy with languid sentiment." Pepys characterized Lely as "a mighty proud man and full of state."

Cartoons. Mantegna's "Triumph of Julius Caesar." Mrs. Caroline M. S. Kirkland had written in *Holidays Abroad.* "The cartoons make one melancholy, for nothing can be more suggestive of the transitoriness of earth's great doings than the pale and scarce distinguishable aspect of these treasures of art." In addition, if they were "not well disposed for light," Melville must have found them disappointing at least. They have since been restored.

Monday, November 12

Redburn. In a letter to Richard Bentley, "New York June 5th 1849," Melville discusses *Redburn,* upon which he is still at work, and proposes it for English publication. It is interesting for what it says of *Redburn;* even more so, of *Mardi.* I give it in full, followed by his further letter of July twentieth, 1849 on the same subject.

Dear Sir — The critics on your side of the water seem to have fired quite a broadside into "Mardi"; but it was not altogether unexpected. In fact the book is of a nature to attract compliments of that sort from some quarters; and as you may be aware yourself, it is judged only as a work meant to entertain. And I cannot but think that its having been brought out in England in the ordinary novel form must have led to the disappointment of many readers, who would have been better pleased with it, perhaps, had they taken it up in the first place for what it really is. — Besides, the peculiar thoughts & fancies of a Yankee upon politics & other matters could hardly be presumed to delight that class of gentlemen who conduct your leading journals; while the metaphysical ingredients (for want of a better term) of the book, must of course repel some of those who read simply for amusement. — However, it will reach those for whom it is intended; and I have already received assurances that "Mardi," in its larger purposes, has not been written in vain.

You may think, in your own mind that a man is unwise, — indiscreet, to write a work of that kind, when he might have written one perhaps, calculated merely to please the general reader, & not provoke attack, however masqued in an affectation of indifference or contempt. But some of us scribblers, My Dear Sir, always have a certain something unmanagable in us, that bids us do this or that, and be done it must — hit or miss.

I have now in preparation a thing of a widely different cast from "Mardi": — a plain, straightforward, amusing narrative of personal ex-

perience — the son of a gentleman on his first voyage to sea as a sailor — no
metaphysics, no cosmic[?] sections, nothing but cakes & ale. I have shifted
my ground from the South Seas to a different quarter of the globe — nearer
home — and what I write I have almost wholly picked up by my own observation under *comical* circumstances. In size the book will be perhaps
a fraction smaller than "Typee"; will be printed here by Harpers, & ready
for them two or three months hence, or before. I value the English copyright at one hundred & fifty pounds, and think it would be wise to put it
forth in a manner, admitting of a popular circulation.

Write me if you please at your earliest leisure; and as you have not yet
sent me any copies of your editions of "Mardi" — (which of course I impute
to the fact of the prodigious demand for the book with you) — I will thank
you to forward me three copies. A note dropped to my friend Mr. Brod-
head of the Legation, will be the means of informing you whether he can
send them to me in the Despatch Bag. If he cannot, the parcel would reach
me by Hamden's Express, — addressed to care of Allan Melville No. 14 Wall
Street, New York.

Very Faithfully, Dear Sir,
Herman Melville

Richard Bentley, Esq.
New Burlington Street

Mr. Bentley replied promptly: Melville wrote again —

New York, July 20th '49

Dear Sir — I am indebted to you for yours of the 20th June. — Your report
concerning "Mardi" was pretty much as I expected; but you know perhaps
that there are goodly harvests which ripen late, especially when the grain
is remarkably strong. At any rate, Mr. Bentley, let us by all means lay this
flattering unction to our souls, since it is so grateful a prospect to you as a
publisher, & to me as an author. But I need not assure you how deeply I
regret that, for any period, you should find this venture of "Mardi" an unprofitable thing for you; & I should feel still more grieved, did I suppose it
was going to eventuate in a positive loss to you. But this cannot be in the
end. — However, these considerations — all, solely with respect to yourself —
prevail upon me to accept your amendment to my overtures concerning my
new work: — which amendment, I understand to be this — £100 down on
the receipt of the sheets, on account of half profits; & that you shall be
enabled to publish a few days previous to the appearance of the book in
America — and this I hereby guarantee.

The work is now going thro' the press, and I think I shall be able
to send it to you in the course of three weeks or so. It will readily make

two volumes got up in your style, as I have enlarged it somewhat to the size of "Omoo" — perhaps it may be a trifle larger.

Notwithstanding that recent decision of your courts of law, I can hardly imagine that it will occasion any serious infringement of any rights you have in any American book. And ere long, doubtless, we shall have something of an international law — so much desired by all American writers — which shall settle this matter upon the basis of justice. The only marvel is, that it does not now exist.

The copies of "Mardi" have not yet come to hand, tho' I sent to the Hamden & Co, to inquire.

Yours sincerely
H. Melville

Richard Bentley, Esq.
New Burlington Street

These letters are in the collection of Mr. Bradley Martin of New York, who kindly permits me to use them.

Copyright question. The lack of an international copyright law made the piracy of American books in England and of English books in America a not uncommon occurrence, if the authors were already well known. The *New York Herald* for November 15, November 23, and December 8, 1849, carried long, indignant editorials on the injustice of the system, or lack of system — the "bold, inexcusable, and cruel robbery of foreign authors" and "the cruel outrage . . . perpetrated on our own writers at home." And *The Home Journal,* January 12, 1850, under the heading *Light Touchings,* brought to its readers the particular effect of this international lawlessness on one of their own New York authors.

— Our friend Herman Melville is one of the first and most signal realizers of the effect of the recent English repudiation of copyright. As our readers probably know, it has been a rule among publishers abroad that an agreement of prior publication, between one of their number and an American author, should be as valid as the legal copyright of an English author. To punish us for our wholesale thieving of English books, they have broken up this protection, by mutual consent, and, now, an American author can no more sell a book in England than Dickens can sell one here — justly enough! Melville went abroad, about the time that this retaliatory

[111]

system came first into action — but knowing nothing of it, and relying on the proceeds of the English editions of his books, for the means of prolonged travel. He writes us that he has abandoned his more extended plans, with this disappointment, and will return sooner than he expected — but there is one passage of his letter so characteristic, that we cannot forbear giving it to the admirers of Typee and Omoo. He says: — "I very much doubt whether Gabriel enters the portals of Heaven without a fee to Peter the porter — so impossible is it to travel without money. Some people (999 in 1000) are very unaccountably shy about confessing to a want of money, as the reason why they do not do this or that; but, for my part, I think it such a capital clincher of a reason for not doing a thing, that I out with it, at once — for, who can gainsay it? And, what more satisfactory or unanswerable reason can a body give, I should like to know? Besides — though there are numbers of fine fellows, and hearts of blood, in the world, whom Providence hath blessed with purses furlongs in length — yet the class of wealthy people are, in the aggregate, such a mob of gilded dunces, that, not to be wealthy carries with it a certain distinction and nobility."

We can fancy Melville to be perfectly sincere in this submission to his lot, as purses are usually allotted in this world — for, if he has but enough to eat and wear, his genius will do more for him in England than the largest fortune of New-York would do, without it. A counting-house repute, and a counting-house education, with unlimited credit in Wall-st., would make a young man inadmissible in English society, even many degrees below where Melville would be received as an equal. — So, my dear Melville! You can very well afford to let the "dunces" be rich, and we are very well assured that you wrote the disclaimer, above quoted, in willing *abstract* verity — however annoying may have been the temporary foregoings which gave rise to your moralizings.

In 1852 Melville signed a petition to Congress for change in copyright law, with Irving, Putnam, Bryant, et al. — Chronological Table in Willard Thorp's *Herman Melville.*

Mr. Murray. John Murray (1808–1892) publisher of *Murray's Handbooks* of foreign countries, *Lavengro,* and a series of *British Classics;* also a friend of eminent writers.

He was a survivor of the patriarchal age of English publishing, when the publisher endeavoured to associate with the functions of the capitalist the eighteenth century traditions of literary patronage. — DNB

Cigar Divan. Cigar Divan, where "for 1s he [the traveller] has the entrée of a handsome room, a cup of coffee and a cigar, and the use of newspapers, periodicals, chess, &c." — *Murray's Modern London* (1851).

In *The Paradise of Bachelors and the Tartarus of Maids,* a sketch published in *Harper's New Monthly Magazine* for April, 1854, Melville writes, "Templars in modern London! Templars in their red-cross mantles smoking cigars at the Divan! — — — No. The genuine Templar is long since departed."

The Paradise of Bachelors and the Tartarus of Maids may be found in *Billy Budd and Other Prose Pieces.*

Joseph Harper. (1801–1870) One of the Harper Brothers publishing company, better known as Wesley Harper. Melville had further remarked about Mr. Harper, "A regular Yankee," and then deleted it.

Judge and Jury. The channel through which Melville heard of the "Judge and Jury" may have been a letter written by George Tomes to George Duyckinck November 16, 1847. It contains the following passage —

First then, have you visited the "Judge and Jury"? If not, go by all means. It is a "Free and Easy," witty, humorous, self constituted court, and is held at The Garrick's Head, a tavern in the neighborhood of Drury Lane Theatre. But by enquiry, any one can tell you where it is. I have on several occasions been very much entertained at this place. The burlesque is admirable & conducted with great apparent gravity and seriousness. All classes of persons visit this humorous court — from the shopman to the peer. I have never been there without witnessing really a fine display of extemporaneous wit and humour. — NYPL–D

I am indebted to Mr. Jay Leyda for this suggestion.

Nicholson. Renton Nicholson (1809–1861), known as the Lord Chief Baron, established the Judge and Jury Society, where he himself presided.

The trials were humorous, and gave occasion for much real eloquence, brilliant repartee, fluent satire, and not unfrequently for indecent witticism. Nicholson's position as a mock judge was one of the sternest realities of eccentric history. Attorneys, when suing him, addressed him as "my lord." Sheriff's officers, when executing a writ, apologized for the disagreeable duty they were compelled to perform "on the court." — DNB

Tuesday, November 13

The Mannings. Marie Manning (1821–1849), Swiss-born wife of George Manning, was hung together with him for a particularly sordid murder. Dickens, who was also present at the hanging, wrote, "I believe the wickedness and levity of the immense crowd collected at the execution could be imagined by no man, and could be presented in no heathen land under the sun." This agrees with the conviction expressed by Melville in *Typee,* chapter XVII, "Comparative Wickedness of Civilized and Unenlightened People."

Illustrative of the intense popular interest in the Mannings is an advertisement carried by *The Times,* London, as early as November 30 of that year.

> Maria Manning, George Manning, Bloomfield Rush, taken from life at their trials, a cast in plaster of Mr. O'Conor, with a plan of the kitchen where he was murdered, models of Stanfield Hall and Potash Farm, and now added to the Chamber of Horrors, at Madame Tussaud and Sons Exhibition, Bazaar, Baker Street, Portman Square. Open from 11 till dark, and from 7 till 10. Admittance 1s; small room 6d extra.

The New York Herald also gave the subject sensational treatment, and quoted Dickens' letter in the December 2 issue.

Incidentally, it is said that Marie Manning suggested to Dickens the character of Mademoiselle Hortense in *Bleak House.*

Wednesday, November 14

Beaumont & Fletcher. Fifty Comedies and Tragedies. Written by Francis Beaumont and John Fletcher, Gentlemen. All in one Volume, Published by the Authors Original copies, the Songs to each Play being added. London. Printed by J. Macock, for John Martyn, Henry Herringman, Richard Marriot, MDCLXXIX.

The volume contains many of Melville's markings and is inscribed: "Herman Melville, London, December, 1849 (New Year's day, at sea) 1850." — HCL–M.

His preference for the large folio form is expressed in two quotations which he wrote in the front of the book:

"I cannot read Beaumont and Fletcher but in folio." — *Charles Lamb.*

P.S. (By the same) "The octavo editions are painful to look at."

Mr. Stibbs. Edward Cambridge Stibbs, bookseller, 331 Strand, from whom Melville bought other books on this visit.

Chapman's Homer. The 1857 editions of *The Odysseys* [sic] *of Homer* and *The Iliads* [sic] *of Homer,* and the 1858 edition of Homer's *Batrachomyomachia* were a present to Melville in November 1858 from George Duyckinck. They contain Melville's markings. — HCL–M

In thanking his friend Melville wrote from Pittsfield, November 6:

> Your gift is very acceptable — could not have been more so. I am glad to have a copy of Chapman's Homer. As for Pope's version (of which I have a copy) I expect it, — when I shall put Chapman beside it — to go off shrieking, like the bankrupt deities in Milton's hymn. — NYPL–D

Colman. George Colman, the elder (1732–1794), playwright, and at one time part owner of Covent Garden Theatre.

> Colman was a man of tact, enterprise and taste; his plays are ingenious and occasionally brilliant. — The characters are as a rule well drawn, and types of eccentricity are well hit off. — Byron contrasted him favourably with Sheridan. — DNB

Mrs. Glover. Julia Glover (1779–1850), was, in her closing days, known as the "Mother of the Stage." She was sometimes considered "a violent actress," but was "generally credited with refinement and distinction." — DNB

Leigh Murray. Henry Leigh Murray (1820–1870), "a painstaking and competent actor, but wanting in robustness, he owed his reputation in part to the ease and naturalness of his style, and to his avoidance of artifice and convention, and to the absence of mannerism." — DNB

[115]

Old Farren. William Farren (1786–1861). "Farren in his later years was the best representative of the present century of old men. Hazlitt said 'He plays the old gentleman [Lord Ogleby], the antiquated beau of the last age, very much after the fashion that we remember to have seen him in our younger days.' He played until his voice grew feeble and his step uncertain." — DNB

Thursday, November 15

Thanksgiving day. A special Thanksgiving day was proclaimed by the Queen to give thanks for the abatement of "the grievous disease with which many places in this kingdom have been lately visited," an epidemic of cholera.

Angel Inn. A famous coach center, depicted by Hogarth in "The Stage Coach or Country Inn Yard," 1747.

glorious chop. This is only one of many instances when Melville voices a succulent appreciation of good food, an appreciation he kept through life. A little book with a big title, inscribed "Lizzie from Herman, Boston, July — 1854" evinces his interest and a desire to share its possibilities with his wife. The comprehensive title is

The Modern Housewife's Receipt Book. A Guide To All Matters Connected With Household Economy. By Mrs. Pullan. Authoress of "The Lady's Library," etc. With Receipts Tested By John Sayer. Esq. Professed Cook of Manchester. The Medical And Other Portions Of The Work Revised By J. Baxter Langley, Esq., Surgeon, Etc., Lately Medical Referee to the "Family Friend." London. Aird and Hutton, 340, Strand: And All Booksellers 1854.

It contains receipts (many of them elaborate and complicated far beyond the glories of a chop) from "Boar's head sauce" to "Blacking," and "directions To Cure Nightmare." — HCL–M

Friday, November 16

big arm and foot. These two fragments must be taken separately. I am indebted to the courtesy of Mr. F. C. Francis of the British Museum for this information.

The keeper of the Department of Egyptian and Assyrian Antiquities has no doubt that this is the left arm of the colossal statue of an Egyptian king, still exhibited in the Egyptian Gallery. "It was found by Belzoni in 1817 at Karnak, near the remains of a granite building of Thothmes III, and for this reason has been regarded as the head of a statue of Thothmes III" (I quote from the British Museum *Guide to the Egyptian Galleries, Sculpture, 1909*).

In the 1848 *Synopsis of the contents of the British Museum* — [is a] "foot from the statue of a God or King." — but nobody in the department has seen the object and cannot now trace it! It is probably that of the archaic Apollo from Delos. "Fragment of a foot of a colossal statue of Apollo — the fragment consists of parts of the four greater toes of the left foot." Found at Delos 1818. (See *A Catalogue of Sculpture in the Department of Greek and Roman Antiquities, British Museum, 1892*. Vol. I, p. 68).

my book. Could it be that Mr. Murray (John Murray III) who was a young man of stability, averse to romances and poetry, when he became the head of the famous publishing house in 1843, felt he had been deceived in the character of *Typee* and *Omoo,* which he had published and had included in his Home and Colonial Library? The Library was an idea of Gladstone's, proposed to John Murray II, to "allow the public to obtain possession of new and popular works at moderate prices." John Murray III acted upon the idea of a "series of useful and entertaining volumes (at 2s. 6d. each) which would contain nothing offensive to morals or good taste, and would appeal, it was hoped, to heads of families, clergymen, school-teachers and employers of labor."

Sir Walter Farquhar had stated in a letter to Lord Ashley (later Lord Shaftsbury) that *Typee* and *Omoo* "are not works that any mother would like to see in the hands of her daughters, and as such are not suited to lie on the drawing-room table." He further urged Lord Ashley to "elicit from him [Murray] some assurance that there shall not appear in his series another volume similar in character, for without such assurance I shall be reluctantly compelled to cease subscribing to the series —."

Such powerful pressure may have ended his professional connection with Melville. It seems to be based on feelings to which he himself might not have been unsympathetic.

[117]

The letter of Sir Walter Farquhar to Lord Ashley, published in *At John Murray's* by George Paston, was brought to my attention by Mr. Jay Leyda.

Nollikens. Joseph Nollikens (1737–1823), sculptor, especially of portrait busts and monuments.

One of Colley Cibber's comedies. Colley Cibber (1671–1757), actor and dramatist. *She Would and She Would Not* was one of his most successful comedies. "He was a sparkling and successful dramatist, a comedian of high mark, a singularly capable and judicious manager, upon whom, to a certain extent, Garrick is said to have modelled himself, and an unequalled critic of theatrical performances." — DNB

Mr. A. Younge. Actor, born in 1806, and author of one play, "The First of May." *Manuscript List of Actors* in the Harvard Theatre Collection; and *History of Early Nineteenth Century Drama, 1800–1850,* by Allardyce Nicoll (1930).

Cock Tavern. "The most personal relic of Tennyson's life in London is the old Cock Tavern on Fleet Street, at the end of Chancery Lane." *Tennyson's London,* by Oliver Huckel (1914).

Saturday, November 17

Murillos.

His whole figure was free, fine, and indolent; he was such a boy as might have ripened into life in a Neapolitan vineyard; such a boy as gypsies steal in infancy; such a boy as Murillo often painted, when he went among the poor and outcast, for subjects wherewith to captivate the eyes of rank and wealth; such a boy, as only Andalusian beggars are, full of poetry, gushing from every rent. — *Redburn.*

Mr. Colbourn. Henry Colbourn, bookseller and publisher, 13 Great Marlborough Street.

Davenant. This copy is now in the collection of Dr. Edward Rhodebeck of New York.

Hudibras. This book was printed in London in 1710 for John Baker at the Black Boy in Pater-noster-Row. On the flyleaf in the upper right-hand corner, in ink, is the signature of L. Duval, and under it the signature of R. Miles, 1765. In the center of the page in Herman Melville's hand there is the following:

Evert Duyckinck
from H. M.

Feb. 2d 1850

85 years after Mr. Miles the old Englishman, in silk small clothes, bought [?] the book at some stall — *you* own it now — who will own it next?

Melville gave this volume to Evert Duyckinck (NYPL–D), with the 1849 three-volume English edition of *Mardi*, recently published. In a letter written in New York Saturday evening, February 2, 1850, Melville says,

> I hurry to follow it up [*Mardi*] with a fine old spicy duodecimo mouthful in the shape of "Hudibras" which I got particularly for yourself at Stribbs's [sic] in the Strand — & [I] a little marvel that your brother George overlooked so enticing a little volume during his rummagings in the same shop. — Pray glance at the title page, & tell me, if you can, what "Black Boy" that was in Pater-noster Row. My curiosity is excited, and indeed aggravated & exacerbated about that young negro. Did the late Mr. Baker have a small *live* Nubian standing at his shop door, like the mocassined Indian of our Bowery tobacconist? I readily see the propriety of the Indian — but in that "Black Boy" I perceive no possible affinity to books — unless, by the way, Mr. Baker dealt altogether in black-letter, — Thomas the Rhymer, Lydgate, & Battle Abbey Directories. — Are they not delicious, & full flavored with suggestiveness, these old fashioned London imprints? — NYPL–D

In a postscript, he adds,

> I return, with my best thanks, to your brother, *three* of the books he loaned me. I can not account for "Cruchley's" accident in the back. — The Guide books for Northern & Central Italy are neither stolen, sold, or mislaid. I will, I think, satisfactorily account for them when I see your brother. They are safe.

Also Willard Thorp's *Herman Melville*. (Willard Thorp notes that "George Frederick Cruchley's *Picture of London* was a standard guidebook of the day.")

Melville seems to have become familiar with *Hudibras* through his association with "Doctor Long Ghost," who "quoted Virgil, and talked of Hobbes of Malmsbury, besides repeating poetry by the canto, especially 'Hudibras.' " *Omoo.*

Sunday, November 18

Rainbow tavern (Tennyson's). "Generally he would stay at the Temple or in Lincoln's Inn Fields; dining at The Cock and other taverns." — Hallam, *Alfred Lord Tennyson A Memoir by his Son.*

Melville seems to have shown a surprising interest in Tennyson associations, at least so far as taverns are concerned. Tennyson is quoted by Edward Fitzgerald as having said, "All fine-natured men know what is good to eat." — *Ibid.*

A part of the interior of the Rainbow Tavern dates back more than a couple of centuries. The chief interest of the Rainbow, however, lies in the fact that it was at first a coffee-house, and one of the earliest in London. It was opened in 1657 by a barber named James Farr. — Henry C. Shelley, *Inns and Taverns of Old London* (1909).

The neighboring tavern keepers indicted him for making and selling "a strong drink called coffee, which annoyed the neighborhood by its evil smell." — Edward Callow, *Old London Taverns,* (1899).

Monday, November 19

Thomas Delf, export bookseller, 16 Little Britain Street.

Duyckinck's. Evert Duyckinck.

Young D's. George Long Duyckinck.

J. M. Langford. The Mr. Langford of 3 Furnival Inn, listed at the end of the *Journal,* with whom he had supper two days later. In 1856 he wrote Mr. Langford a letter of introduction for his brother-in-law, Samuel Shaw.

Mrs. Welford's. Probably the wife of Charles Welford, bookseller, 7 Astor House, New York, who wrote the catalogue for the sale of Charles Lambs' books, which was published in the *Literary World,* February 5, 1848. She was the daughter of Mrs. Daniel, on whom Melville called on December 14, and again on December 16.

Albert Smith. Albert Richard Smith (1816-1860), author, lecturer and entertainer, whom Melville met two days later. *The Natural History of the Ballet Girl. By Albert Smith. Illustrated by A. Hemming, London: D. Bogue, 86 Fleet Street, MDCCCXLVII,* is inscribed "Lizzie from Herman. June 13th, 1886, N.Y." This is one of a set of four small volumes, which includes also Albert Smith's *Natural History of the Gent* (1847), A. B. Beach's *London on the Thames* [18—?] and J. W. Carleton's *Natural History of the "Hawk" tribe* [1848?] — HCL–M

the Longmans. Thomas Longman (1804–1879) and William Longman (1813–1877), third generation of publishing Longmans. The former had become head of the firm in 1842, the latter a partner in 1839. The firm's best known publications were Lord Macaulay's works and an edition of the New Testament illustrated with wood engravings of paintings by the great masters. William, an ardent Alpine mountain climber, was noted for his courtesy to men of letters and to his brethren of "the trade." The Longman Melville met was "very polite."

Lord John Manners, (1818–1906), second son of the fifth Duke of Rutland; M. P., and holder of various public offices. He is the Lord Henry Sidney in Disraeli's *Coningsby. — Complete Peerage of England, Scotland, Ireland, Great Britain and the United Kingdom. Extant, Extinct, or Dormant,* alphabetically arranged and edited by G. E. C. (London, 1893).

Monckton Milnes. Richard Monckton Milnes (1809–1885) literary dilettante of liberal views, social tastes, and romantic generosity. He became Lord Houghton, editor of Keats.

Mr. Milnes is a very agreeable, kindly man, resembling Longfellow a good deal in personal appearance; and he promotes, by his genial manners, the same pleasant intercourse which is so easily established with Longfellow. . . . He is considered one of the best conversationalists at present in society: it may be so; his style of talking being very simple and natural, anything but obtrusive, so that you might enjoy its agreeableness without suspecting it. . . . Mr. Milnes told me that he owns the land in Yorkshire, whence some of the pilgrims of the Mayflower emigrated to Plymouth, and that

Elder Brewster was the Postmaster of the village. He takes pride in the ownership. . . . Nathaniel Hawthorne in his *Journal*, September 22, 1854. From *The Heart of Hawthorne's Journals*, edited by Newton Arvin (Boston, 1929).

Lady E. Drummond. Elizabeth Frederica (1802–1886) married Andrew Robert Drummond of Cadland March 7, 1821. She was the older sister of Lady Emmeline and Lord John Manners. She lived at 2 Bryanstone Square. — *Burke's Peerage Baronetage & Knightage* (1938).

Willis. Nathaniel P. Willis (1806–1867), New York journalist, poet, editor, and dramatist. His work as interpreted by Lowell was "not deep as a river, but who'd have it deep?"

M. F. Tupper. Martin Farquhar Tupper (1810–1889), author, among other works, of the immensely popular *Proverbial Philosophy*, which ran through many editions in London, New York, Philadelphia, and Buffalo from 1845 to 1854. It was "one of the most successful candidates . . . for the Drawing-room table," in spite of what its detractors dubbed its "muffin morality" — or perhaps *because* of that suggestive quality. His name appears in the address list at the end of the journal.

According to Mr. Jay Leyda, Melville had ordered on September 8, 1846 a copy of *Proverbial Philosophy* from Wiley and Putnam. It is not known what he thought of it.

sister. Lady Emmeline Charlotte Elizabeth Stuart Wortley, second daughter of John Henry, Fifth Duke of Rutland, distinguished as a poet. She married Honorable Charles Stuart Wortley in February, 1831. — *Burke's Peerage Baronetage & Knightage* (1938).

Macready, William Charles (1793–1873). He "considered that to be a great actor it was advisable for him to become a good scholar, an accomplished gentleman, a well-ordered man, with a well-regulated mind, and finely cultivated taste." Though he lacked Kean's passion, he stood very high in the opinion of such critics as Hazlitt, Leigh Hunt, and Talfourd. — DNB

Panted. Alternative reading, "painted" (as Othello).

James Wallack. James William Wallack (1791?–1864), actor from childhood of many and varied parts. He was son, brother, husband, and father to actors. He was said to be "indifferent in tragedy, admirable in melodrama, and always pleasing and delightful in light comedy, in which, however, the spectator was always sensible of a hidden want." He played much in America, crossing and recrossing the Atlantic, and he was the founder of Wallack's Theatre, New York. — DNB

Miss Reynolds. Jane Louisa Reynolds (1824–1907), wife of Baron Brampton, both actress and singer. She made her debut at the Haymarket Theatre, September 7, 1846 as Kate O'Brien in a comedy called *Perfection.*

Horrible Roderigo. Charles Selby (1802?–1863) was the Roderigo in this performance of *Othello.* He was both actor and dramatist. His plays were of "the lightest description," and he was considered "a useful and responsible actor" with a face that had "a quaint comic twist." — DNB

I am indebted to Dr. William B. Van Lennep, Curator of Harvard's Theatre Collection for identifying the Roderigo of this performance.

Buckstone. John Baldwin Buckstone (1802–1879), actor and dramatist, famous for the broad humor of his best acting and the authorship of between one and two hundred successful plays.

Tuesday, November 20

letters of introduction. Though Melville went supplied with many letters, there is one in particular that he would have liked to have and which there is no record, I believe, of his having received. He had written his father-in-law,

Monday Sept 10th [1849]

• My dear Sir — In writing you the other day concerning the letters of introduction, I forgot to say, that could you conveniently procure me one from Mr. Emerson to Mr. Carlyle, I should be obliged to you. — We were concerned to hear that your [*sic*] were not entirely well, some days ago; but I hope you will bring the intelligence of your better health along with you, when you come here on that promised visit, upon which you set out the day after tomorrow. Lizzie is most anxiously expecting you — but Malcolm seems to await the event with the utmost philosophy. — The weather here at present is exceedingly agreeable — quite cool, & in the morning, bracing.

My best remembrances to Mrs. Shaw & all.

Most Sincerely Yours
H. Melville

If, besides a letter to Mr. Carlyle, Mr. Emerson could give you *other* letters, I should be pleased. The Board of Health have ceased making reports. — the Cholera being almost entirely departed from the city.

Judge Lemuel Shaw (1781–1861), his father-in-law, liberal and incorruptible Chief Justice of the Supreme Court of Massachusetts from 1830 to 1860, to whom Melville dedicated *Typee*. "Gratitude" is expressed in the dedication, "sincere respect" in the presentation copy, now in my possession.

Mr. Rogers. Samuel Rogers (1763–1855), author of *Pleasures of Memory*, man of fortune with literary and artistic tastes, generous of advice and money to authors and artists, renowned for his breakfasts and table-talk. Fanny Kemble said of him, "He certainly had the kindest heart and unkindest tongue of anyone I ever knew." His *Pleasures of Memory* "may be regarded as the last embodiment of the poetic diction of the 18th century. Here is carried to the extremest pitch the theory of elevating and refining familiar themes by abstract treatment and lofty imagery."

Melville's father-in-law, Judge Shaw, had asked Edward Everett to furnish Melville with the letter of introduction to Mr. Rogers which follows.

Cambridge: 3rd Sept., 1849

My dear Mr. Rogers, — It is such an age since I have written to you, that I am really under obligation to my honored friend the Chief Justice of Mass. who has asked me for his son-in-law Mr. Melville, the favor of one or two letters to London. This gentleman (I am sorry to say) is not, known to me personally. He is known to you and the entire reading world by his "Typee" and "Omoo," and another work of the same class, which I have not yet seen. I understand Mr. Melville's character to be altogether such as warrants me in commending him to your kind notice. His brother, who was Secretary of Legation under Mr. M'Lane, was, I think, known to you. Few of our writers have been as successful at home as Mr. Herman Melville, and I am happy to perceive that his productions are well known on your side of the water.

Mr. Melville is going to pass a few months in England and France, and while he is in London I want him to see a few of those choicest spirits, who even at the present day increase the pride which we feel in speaking the language of Shakespeare and Milton. In a word, my dear friend, I want you to admit him to the freedom of No. 22 St. James's Place.

I need not tell you how constantly we think of you: — how often we speak of you: — how regularly we do the honors of your portrait to all who come to us. I should be delighted to have under your hand & seal, the confirmation of the good accounts I have of you from others; and I pray you to believe me, my dear Mr. Rogers, with the Strongest Attachment.

Sincerely yours,
Edward Everett.

An unsigned transcription of this letter is among Edward Everett papers. — MHS–E

Mr. Lawrence. Abbott Lawrence (1792–1855), Boston merchant, manufacturer, diplomat, statesman, and philanthropist. He was minister to Great Britain for three years under President Taylor.

Duke of Rutland. John Henry (Manners), fifth Duke of Rutland (1778–1857), held various public offices. He is "the Duke" in Disraeli's *Coningsby.* "Of tall and noble presence, exceedingly elegant and dignified in manner, but singularly courteous in his reception of those who had business with him." This description of him in the *Gentleman's Magazine,* 1857, is included in the *Complete Peerage* (1893).

"Lord Henry Sidney's father, the Duke, and his home Beaumanoir, so clearly point to the Duke of Rutland and Belvoir Castle, that no contemporary could have hesitated in taking him for Lord John Manners." — Introduction by B. M. Langdon-Davies to Everyman's Library edition of Disraeli's *Coningsby* (1928).

Russell Sturgis. Shipping merchant and commissioner of pilots, New York. Descendant of Edward Sturgis (1634), one of the first settlers of Charlestown, Massachusetts, and father of Russell Sturgis, architect, critic, and writer. Melville met him four days later.

Joshua Bates (1788–1864), financier and philanthropist, an American living most of his life in England. Founder in 1852 of the Boston Public Library. Melville met him four days later.

Moxon. Edward Moxon (1801–1858), distinguished publisher of Shelley, Wordsworth, Tennyson, Lamb, etc. He was an intimate friend of Lamb's and husband of his adopted daughter.

In a letter of Melville's to Richard Henry Dana, Jr., New York, May 1, 1850, he elaborated his *Journal's* description of the meeting with Mr. Moxon.

Let me mention to you now my adventure with the letter you furnished me to Mr. Moxon. Upon this, as upon other similar occasions, I chose to wave cerimony [sic]; and so arranged it, that I saw Mr. Moxon, immediately after the reception of the letter. — I was ushered into one of those jealous, guarded sanctums, in which these London publishers retreat from the public gaze. It was a small, dim, religious looking room — a very chapel to enter. Upon the coldest day you would have taken off your hat in that room, tho' there were no fire, no occupant, & you a Quaker. — You have heard, I dare say, of that Greenland whaler discovered near the Pole, adrift & silent in a calm, with the frozen form of a man seated at a desk in the cabin before an ink-stand of icy ink. Just so sat Mr. Moxon in that tranced cabin of his. I bowed to the spectre, & received such a galvanic return, that I thought something of running out for some officer of the Humane Society, & getting a supply of hot water & blankets to resuscitate this melancholy corpse. But knowing the nature of these foggy English, & that they are not altogether impenetrable, I began a sociable talk, and happening to make mention of Charles Lamb, and alluding to the warmth of feeling with which that charming punster is regarded in America, Mr. Moxon lighted up — grew cordial — hearty; — & going into the heart of the matter — told me that he (Lamb) was the best fellow in the world to "get drunk with" (I use his own words) & that he had many a time put him to bed. He concluded by offering to send me a copy of his works (not Moxon's poetry, but Lamb's prose) which I have by me, now. It so happened, that on the passage over, I had found a copy of Lamb in the ship's Library —

& not having previously read him much, I dived into him, and was delighted — as every one must be with such a rare humorist & excellent hearted man. So I was very sincere with Moxon, being fresh from Lamb. — MHS-D

Also, Harrison Hayford, "Two New Letters of Herman Melville," E.L.H., *A Journal of English Literary History*, 11, 1, (March, 1944), which was brought to my attention by Mr. Hayford.

Dana. Richard Henry Dana (1815–1882). Moxon published his *Two Years Before The Mast* in 1840. Dana had furnished Melville with a letter of introduction to Moxon.

Boston, Sep. 12, 1849

My dear Sir,

Allow me to introduce to you my friend, Mr. Herman Melville, of New York. Mr. Melville is the Author of those well known & popular narratives of adventures at sea & in the Pacific Islands, with which you are doubtless familiar. So many persons have affected to believe his a nom de guerre, that I am happy to learn his intention of visiting Europe in propria persona, so that his readers there may satisfy themselves not only that he is a veritable person, but a most agreeable gentleman of one of our best families. In some cases I might hesitate whether my relations with yourself would authorize me to introduce a friend to your attentions, but I am sure you will derive so much gratification from the acquaintance with Mr. Melville, that I feel as if I were only communicating to you a pleasure of my own.

I should like very much to have him see Capt. Jones, if he is in London.

Most sincerely yours
Richard H. Dana, Jr.

Edward Moxon, Esquire.

Melville had acknowledged this just before he sailed.

New York, Oct. 6th, 1849

My Dear Mr. Dana — If I have till now deferred answering your very kind letter by Judge Shaw, it has been only, that I might give additional emphasis to my reply, by leaving it to the eve of my departure. Your letter to Mr. Moxon is most welcome. From his connection with Lamb, & what I have chanced to hear of his personal character, he must be a very desirable acquaintance. — Your hint concerning a man-of-war has, in anticipation, been acted on. A printed copy of the book is before me. As it will not appear for some two or three months, may I beg of you, that you will consider this communication confidential? The reason is obvious.

[127]

This man-of-war book, My Dear Sir, is in some parts rather man-of-*warish* in style — rather aggressive I fear. — But you, who like myself, have experienced in person the usages to which a sailor is subjected, will not wonder, perhaps, at anything in the book. Would to God, that every man who shall read it, had been before the mast in an armed ship, that he might know something himself of what he shall only read of. — I shall be away, in all probability, for some months after the publication of the book. If it is taken hold of in an unfair or ignorant way; & if you should possibly think, that from your peculiar experiences in sea-life, you would be able to say a word to the purpose — may I hope that you will do so, if you can spare the time, & are generous enough to bestow the trouble? — Your name would do a very great deal; but if you choose to keep that out of sight in the matter, well & good. — Be not alarmed, — I do not mean to bore you with a request to do anything in this thing — only this: If you feel so inclined, do it, & God bless you.

Accept my best thanks for your kindness & believe me fraternally yours — a sea-brother —

H. Melville

Richard H. Dana, Jr., Esq.

A little nursery tale of mine (which possibly, you may have seen advertised as in press) called "Redburn" is not the book to which I refer above.
— MHS–D

In a second letter to Dana, New York, May 1, 1850, Melville writes of

those strange, congenial feelings, with which after my first voyage, I for the first time read "Two Years Before the Mast," and while so engaged was, as it were, tied & welded to you by a sort of Siamese link of affectionate sympathy — that these feelings should be reciprocated by you, in your turn, and be called out by any White Jackets or Redburns of mine — this is indeed delightful to me. — MHS–D

The two letters to Dana were edited and published in ELH, *A Journal of English Literary History*, vol. II, No. 1 (March, 1944), by Harrison Hayford.

I am indebted to Mr. Henry W. L. Dana for a copy of Richard Dana's letter of introduction, and to Mr. Hayford for a reprint of his article containing Melville's two letters.

But his wife . . . . crawling things. This passage had been deleted by Melville. See note under Tuesday, November 27, for Melville's reversal of judgment.

Lady Bulwer. Probably the wife of the able and popular Sir Henry Bulwer, who had just been appointed ambassador at Washington in April.

Wednesday, November 21

The negro. Whether Melville actually had a talk with this negro or not is a question; but long years after, he writes of a tale of certain "sanctioned irregularities" in the British Navy of Nelson's time, "communicated to me now more than forty years ago by an old pensioner in a cocked hat, with whom I had a most interesting talk on the terrace at Greenwich, a Baltimore negro, a Trafalgar man." *Billy Budd,* in *Billy Budd and Other Prose Pieces.*

Charles II. In Charles II's time Greenwich Hospital was still Greenwich Palace. It became Greenwich Hospital when King William III and Queen Mary gave it to the Navy as a hostel for pensioners, which it remained until 1873, when it became the Naval College. "There is no reliable personal story about King Charles II at Greenwich and there is nothing in the archives to that effect. There is however a tale, which is not believed by the historians, that Nell Gwynne used a certain room in the College."

I am indebted for this information to the courtesy of Vice Admiral Sir E. J. Patrick Brind, of the Royal Naval College, Greenwich.

good fellow. In a letter to Evert Duyckinck, Paris, December 2, 1849, Melville wrote:

> Give my best remembrances to your brother. Tell him I stumbled upon an acquaintance of his — a book dealer in the Strand. [Edward C. Stibbs] Tell him that Davidson proved a good fellow, & that we took some punch together at the Blue Posts. — NYPL–D, also Willard Thorp, *Herman Melville.*

Tom Taylor (1817–1880), member of the staff and finally editor of *Punch,* and prolific writer of popular plays.

> Mr. [Tom] Taylor is reckoned a brilliant conversationalist; but I suppose he requires somebody to draw him out and assist him; for I could see nothing that I thought very remarkable on this occasion. He is not a kind

of man whom I can talk with, or greatly help to talk; so, though I sat next to him, nothing came of it. He told me some stories of his life in the Temple — little funny incidents, that he afterwards wrought into his dramas; in short, a sensible, active-minded, clearly perceptive man, with a humourous way of showing up men and matters, but without originality, or much imagination, or dance of fancy. — Nathaniel Hawthorne in his *Journal,* April 9, 1856. From *The Heart of Hawthorne's Journals,* edited by Newton Arvin (Boston, 1929).

2 A.M. Melville evidently had such a pleasant time at this prolonged supper party, and remembered such agreeable traits in his host, that seven years later he gave his brother-in-law, Samuel Shaw, the following letter of introduction to Mr. Langford:

Dear Sir: — Allow me to introduce to you Mr. Samuel Shaw of Boston, Massachusetts, — son of Chief Justice Shaw of that state — who, being on a European tour, proposes to spend some time in London.

He is of that temper and those tastes, which I am sure, will not prove uncongenial to you and your friends; while from your acquaintance, he could not fail to reap, as a traveller, both pleasure and profit.

Whatever you may be able to do for him in the way of directing his attention to interesting objects or persons in London, will be gratefully remembered by

His brother-in-law
Yours Very Truly
Herman Melville
Pittsfield, Mass. April 5th, 1856

Mr. Langford
Furnival's Inn.

— HCL–M

Lamb. Among the Lamb items was Thomas Noon Talfourd's *Final Memorials of Lamb.* At the top of the title page it bears the inscription, "To Herman Melville, Esq., with the Publisher's regards."

In the lower left hand corner of the title page Melville wrote, "London. Nov 21, 1849." The four following quotations are among the numerous passages he marked.

Few things irritated him [Lamb] more than the claims set up for the present generation to be wiser and better than those which have gone before it. . . . (p. 169).

[130]

— the apparent anomalies of Mr. Godwin's intellectual history arose from the application of his power to the passions, the interests, and the hopes of mankind, at a time when they enkindled into frightful action, and when he calmly worked out his problems among their burning elements with the "ice-brook's temper," and the severest logic. . . . (p. 142).

— of all celebrated persons I ever saw, Coleridge alone surpassed the expectation created by his writings; for he not only was, but appeared to be, greater than the noblest things he had written. . . . (p. 196).

— the want of that Imagination which brings all things into one, tinges all our thoughts and sympathies with one hue, and rejects every ornament which does not heighten or prolong the feeling which it seeks to embody. . . . (p. 162). — Library of Princeton University.

Thursday, November 22

Rowland's Kalydore. Popular specific for "improving and beautifying the Skin and Complexion." — Advertisement, *Murray's Handbook for London 1851.*

Friday, November 23

Bogue. David Bogue, publisher and print seller. Among his publications were Albert Smith's *The Natural History of the Gent,* the first of a series of social natural histories begun in 1847, which also included *The Natural History of the Ballet Girl.* See note under Monday, November 19, *Albert Smith.*

figs. Melville must have been fond of figs all his life. One of my own earliest remembrances of his room in Twenty-Sixth Street, New York, was the bag of figs on his paper-piled table. See Raymond Weaver's *Herman Melville, Mariner and Mystic,* p. 379.

Lockhart. John Gibson Lockhart (1794–1854), son-in-law, intimate friend, and biographer of Scott. "Lockhart was proud and reserved, and gave an impression of coldness in general society. But he could relax among intimate friends, and had the rare charm which accompanies the occasional revelation under such circumstances of a fine mind and character." Melville evidently did not meet him in the right circumstances. Edward Everett refers to this different Lockhart in a letter to Lady Ashburton, given in a note on Lord Ashburton, November 24 of the *Journal.*

[131]

Cooke. Robert Francis Cooke, cousin and partner of John Murray, with whom Melville dined later.

At dinner . . . . somehow, by. This passage had been deleted by Melville.

Conversing. This word was changed to *conversed* after the cancelled passage.

Dr. Holland. The eminent, much-travelled, socially popular Dr. Henry Holland (1783–1873) who in the early forties had attended the brother of Edward Everett in China. He would have commended himself to Melville for other reasons than his knowledge of far places of the earth. In middle life he had decided not to allow his professional income to exceed £5000, and to use his leisure in study, recreation, and travel. In his later years he retired from practice, but continued to make long tours.

Such is . . . . upon it had been deleted.

Saturday, November 24

Mr. Chapman. Frederic Chapman (1823–1895), later head of the firm, publisher of Dickens, the Brownings, George Meredith, etc., and founder of the *Fortnightly Review.*

Bohn's. Henry George Bohn (1796–1884), independent, energetic publisher of sure-fire successes.

Dolly. A nickname for his wife. Also the name of the ship which takes the narrator of *Typee* to the Marquesas.

Lord Ashburton. This probably refers to the first Lord Ashburton who had married an American wife, Anne Louisa Bingham of Philadelphia, and who had died the year before. Lord and Lady Ashburton were among the friends of Edward Everett, with whom he corresponded after his departure from England.

I quote from a letter of his to Lady Ashburton, Cambridge, Massachusetts, November 15, 1847 for its several allusions pertinent to this journal.

. . . I live in a round of petty engagements, which fill up the time and engross my thoughts. But this morning I am determined to have a holiday, and to pass it at the "Grange"; to meet the cheerful circle in the breakfast room at ten, (here our hour at this season is 1/2 past 7) listen with mute satisfaction to the encounter of wits, between Lady Darcy & Mr. Lockhart; glance at the daily Phillipic of the Newspapers, which confirm to us the pleasing intelligence, repeated every morning for the last forty years, that the world is at length certainly coming to an end; join Lord A. in a three hours' walk, in which if the weather be decent, we shall perhaps be honored with the company of those who "make a sunshine in the shady place": — and returning fill up the day with that alternative of retirement & society; reading & talk; which makes country life in England the most delightful existence in the world. . . . [News of the Mexican war]. . . . Meantime, & to offset these present troubles, let me send you a tale of the woes of other times; a pretty little poetical romance called "Evangeline" by Mr. Longfellow. I have not read it all; my ladies, (who desire their kindest remembrance) are delighted with it. You have, I suppose, added Mr. Prescott's "Peru" to your American library, & perhaps "Omoo" by the author of "Typee." I see some of your periodical critics suppose that Mr. Melville is an imaginary personage. Such is not the case. He is a real man; just married in Boston. His brother was Mr. McLane's Secretary of Legation: that his heroes and heroines were also real men and women is more than I would vouch for; not less so, however, I imagine than the fanciful lords & ladies that fill the chapters of modern fashionable novels. — MHS–E

A transcription of this letter is among the Edward Everett papers.

The nephew is possibly Thomas (1799–1873), son of Sir Thomas Baring (brother of the first Baron Ashburton), who had a taste for fine pictures.

Mr. Peabody. George Peabody (1795–1869), merchant, financier, and philanthropist. The Peabody Institutes of Baltimore and of Peabody, Massachusetts, and the Peabody Museums of Harvard and Yale were among his large gifts. He is remembered as kindly and generous; and "though simple in his personal tastes, moved urbanely in London society." In gratitude for his pleasant life in London, he gave large sums to build better houses for the poor of the city.

Chamier. Frederick Chamier (1796–1870), besides writing sea novels, also edited, and continued to 1827, James's *Naval History.*

a strong contrast . . . . dinner last night. This had been deleted. *fat, ugly.* Also deleted in the manuscript.

Gansevoort. Gansevoort Melville (1815–1846), his brother, who was Secretary to the American Legation in London, under Minister McLane, at the time of his death in 1846. He had distinguished himself as an orator in the previous Democratic national campaign. He was a singularly handsome, attractive young man, who probably had more in common intellectually with his brother, Herman, than either of his other brothers. He showed a devoted interest in Herman's literary fortunes in London.

silence.

All profound things, and emotions of things are preceded and attended by Silence. What a silence is that with which the plae bride precedes the responsive *I will,* to the priest's solemn question, *Wilt thou have this man for thy husband?* In silence, too, the wedded hands are clasped. Yea, in silence the child Christ was born into the world. Silence is the general consecration of the universe. Silence is the invisible laying on of the Divine Pontiff's hands upon the world. Silence is at once the most harmless and the most awful thing in all nature. It speaks of the Reserved Forces of Fate. Silence is the only Voice of our God.

Nor is this so august Silence confined to things simply touching or grand. Like the air, Silence permeates all things, and produces its magical power, as well during that peculiar mood which prevails at a solitary traveller's first setting forth on a journey, as at the unimaginable time when before the world was, Silence brooded on the face of the waters. — *Pierre, or, The Ambiguities.*

Sunday, November 25

Moses & Son. Probably Elias Moses and Son, wholesale and retail clothiers, tailors, hatters, 83–86 Aldgate Street. — *Post Office London Directory, 1847.*

"Oh Solitude!"   *Oh Solitude! where are the charms*
            That sages have seen in thy face?
— Verses supposed to be written by Alexander Selkirk, Bartlett, *Familiar Quotations.* Among Melville's books was *The solitudes*

[134]

of nature and of man; or, the loneliness of human life, by William Rounseville Alger (Boston 1867). It contains his marginal markings and comments. William R. Alger (1822–1905), was a public-spirited, independent Unitarian minister of Boston, scholar, and author. — HCL–M

Tuesday, November 27

The Nore. The mouth of the Thames, where the river is six miles wide and the water very choppy. Scene of the "Great Mutiny" of 1797, used by Melville in *Billy Budd* to give strength to the action of authority in the conclusion of the narrative.

"To the Empire, the Nore Mutiny was what a strike in the fire-brigade would be to London threatened by general arson." — *Billy Budd and Other Prose Pieces.* Manuscript of *Billy Budd.* — HCL–M

> *bottle.* Seems to me, putting this and that and the other thing together, it's a sort of alphabet that spells something. Spoon, tumbler, water, sugar, — brandy — that's it. O-t-a-r-d is brandy. Who put these things here? What does it all mean? Don't put sugar here for show, don't put a spoon here for ornament, nor a jug of water. There is only one meaning to it, and that is a very polite invitation from some invisible person to help myself, if I like, to a glass of brandy and sugar, and if I don't like, let it alone.

See *Israel Potter,* Chapter IX, "Israel is initiated into the mysteries of lodging houses in the Latin Quarter." Also, *Mardi* vol. I, a chapter on "Otard."

Mrs. Lawrence . . . . before. This had been deleted.

Alexandrian Manuscript. The Codex Alexandrius in Greek (Fifteenth Century) next to the Codex Sinaiticus (in Petrograd) and the Codex Vatinicus (in Rome) the oldest extant MS of the Bible. — Muirhead's *London and its Environs,* 1919.

> *portmanteau.*

> For all men who say *yes,* lie; and all men who say *no,* — why, they are in the happy condition of judicious, unincumbered travellers in Europe; they cross the frontiers into Eternity with nothing but a carpet-bag, — that is to

say, the Ego. — Letter of Melville's to Hawthorne, April, 1851, given by Julian Hawthorne in *Nathaniel Hawthorne And His Wife* (Boston, 1885).

In November 1856 Hawthorne reported from Southport that Melville

sailed on Tuesday, leaving a trunk behind him, and taking only a carpet-bag to hold all his travelling-gear. This is the next best thing to going naked; and as he wears his beard and mustache, and so needs no dressing-case, — nothing but a toothbrush, — I do not know a more independent personage. He learned his travelling habits by drifting about, all over the South Seas, with no other clothes or equipage than a red flannel shirt and a pair of duck trousers. Yet we seldom see men of less criticisable manners than he. — Nathaniel Hawthorne, *The English Note Books,* edited by Randall Stewart, p. 437.

See also Julian Hawthorne, *Nathaniel Hawthorne And His Wife*; and Raymond Weaver, *Herman Melville, Mariner And Mystic*.

Galignani's. Galignani's Reading Room, 18 Rue Vivienne, had all the best newspapers of the world, a list of all Englishmen living in or visiting Paris, and a circulating library.

statuary. A more than life-size bust of Antinoüs on a pedestal was among Melville's belongings. He must have seen the Antinoüs of the Louvre; but he makes no specific mention of one of his favorite pieces of sculpture till he saw the Vatican Antinoüs in 1857, when he notes

Antinoüs, beautiful. — Walked to the Pincian hill — . . . . Fashion & Rank — Preposterous touring within a stone's throw of Antinoüs. How little influence has truth in the world! — Fashion everywhere ridiculous, but most so in Rome. — *Journal Up the Straits,* published by the Colophon in 1935 with an introduction by Raymond Weaver.

California. The great gold rush by the "Forty-niners."

Arrival of the Steamship Crescent City, with Over a Million of Dollars in Specie and Gold Dust. — Headline under *Important From California New York Herald,* December 8, 1849.

Also

This [*Sherlock's Digging, Mariposa*], *Mr. Editor, is unquestionably the place for big chunks* . . . . — *Intelligence from the Gold Mines.*

See *Mardi*, vol. II, a chapter on "They Encounter Gold Hunters."

"Lit. World." *The Literary World*. Melville was to read the full review at Wiley's in London on December 18.

Saturday, December 1

Melville was frequently uncertain of the date. This date was originally "Friday, 30th Nov." Then "Friday" was cancelled for "Saturday" and the whole was rewritten interlineally.

Cousin. Victor Cousin (1792–1867), French philosopher and lecturer at the Sorbonne, editor of Proclus and Descartes, and translator of Plato.

> Cousin's gifts lay in the direction of observation and generalization rather than analysis or original speculation. He left no distinctive permanent principle of philosophy. But his eclecticism, proceeding as it did from an appreciation of nearly every system of philosophy ancient and modern, was a valuable influence in the direction of toleration and width of view. — *Encyclopaedia Britannica.*

I leave it to the students of philosophy (and of Melville) to speculate on whether he would have been greatly interested in Cousin's views at this period, if he could have understood the language. In 1851, Melville published *Moby Dick*, and in 1852, *Pierre*. In 1854 Cousin published *Du vrai, du beau, et du bien*.

Rachel. Elizabeth Felix (1821–1858). Melville wrote to Evert Duyckinck the following day,

> The other evening I went to see Rachel — & having taken my place in the "que" (how the devil do you spell it?) or tail — & having waited there for a full hour — upon at last arriving at the ticketbox — the woman there closed her little wicket in my face — & so the "tail" was cut off. — NYPL–D

Given in full by Willard Thorp, *Herman Melville* (1938).

Tuesday, December 4

Parnassus. This may refer to a "bronze monument called the *Parnasse Français*, made by Du Tillet in 1778, representing a mountain with statues of poets and authors of the reign of Louis XIV." Murray describes it as "unmeaning." — *Murray's Hand-Book Paris* (1867).

Museum probably refers to the collection of antiquities, or *Cabinet de Médailles et Antiques* in the same building.

Wednesday, December 5

Dupuytren. Medical museum evidently recommended by his friend, Dr. Augustus Gardner. See Introduction. Murray's *Handbook For Visitors to Paris, 1867* lists it among the sights, but as "open to gentlemen only."

Hôtel de Cluny. See his use of this in *Moby Dick.*

> Ahab's larger, darker, deeper part remains unhinted. But vain to popularize profundities, and all truth is profound. Winding far down from within the very heart of this spiked Hotel de Cluny where we here stand — however grand and wonderful, now quit it; — and take your way, ye nobler, sadder souls, to those vast Roman halls of Thermes; where far beneath the fantastic towers of man's upper earth, his roots of grandeur, his whole awful essence sits in bearded state; an antique buried beneath antiquities, and throned on torsos! So with a broken throne, the great gods mock that captive king; so like a Caryatid, he patient sits, upholding on his frozen brow the piled entablatures of ages. Wind ye down there, ye prouder sadder souls! question that proud, sad king! A family likeness! ay, he did beget ye, ye young exiled royalties; and from your grim sire only will the old State-secret come. — *Moby Dick.*

"This is Amadis, son of a king," was written on the parchment hung round the neck of Amadis de Gaul by his mother. See note on Orianna, October 20.

He invited me to dine with him. This was changed from "Invited him to dine with me."

Thursday, December 6

Selah! Melville was at what might be called a transition point when he wrote "Selah."

Of the three letters of introduction which Edward Everett wrote for his friend Judge Shaw's son-in-law, those to Monckton Milnes and Samuel Rogers were used, as we have seen. The third, to M. de Beaumont, Melville does not seem to have used. Assuming that he took it with him, why did he not present it?

Diffidence due to his ignorance of French, possibly; or knowledge that he had but a very short time to stay in the city; or both: two obvious, practical reasons. George Adler suggests another reason in a letter to George Duyckinck, dated Paris, February 16, 1850.

> Our friend Mr. Melville has, I hope, long ago reached his home again safely, and you will have gained from him an account of our voyage and peregrinations in England and London. I regretted his departure very much; but all I could do to check and fix his restless mind for a while at least was of no avail. His loyalty to his friends at home and the instinctive impulse of his imagination to assimilate and perhaps to work up into some beautiful chimaeras (which according to our eloquent lecturer on Plato here [Cousin] constitute the essence of poetry and fiction) the materials he had already gathered in his travels, would not allow him to prolong his stay. — NYPL–D

This corroborates Melville's general statement in a letter to Richard Bentley, New York, June 5, 1849:

> Some of us scribblers — always have a certain something in us, that bids us do this or that, and be done it must — hit or miss.

This letter is in the collection of Mr. Bradley Martin of New York, by whose permission I quote from it.

Because of their interest, I give Edward Everett's letter to Judge Shaw, and his letter of introduction to M. de Beaumont in full. The reference in the latter to a "criticism in the *Revue des deux Mondes*" is to the article, May 14, 1849, "Voyages réels et fantastiques d'Herman Melville," by Philarète Chasles.

> My dear Sir,
> Your letter of the 31 Aug. was handed to me on Saturday Evg.
> I am obliged as a general rule to decline giving letters of Introduction to my friends Abroad, for several reasons which will occur to you. I am however always happy to have it in my power to make exceptions, though I do it but rarely. The extraordinary literary merit of Mr. Melville warrants me in doing so, in this case — and a desire to meet any wish of yours furnishes me a strong additional motive.
> I enclose you for Mr. Melville a letter to Mr. Monckton Milnes, a young gentleman of Fortune; — an M. P.; — A Cambridge man of Trinity College; — A poet of considerable celebrity; — A very companionable person in the highest society; and another to Mr. Nestor Rogers, the Nestor of English

poets now I think in his 87th Year, whose table has been for years the centre of the best society in London.

M. de Beaumont is the grandson-in-law of Lafayette: he travelled in this country in 1833 with M. de Tocqueville to examine our penitentiaries; — he has written an ante [sic]-slavery novel called *Marie ou l'esclavage* & a work on Ireland. He was sent last year by the Provisional government of France as Minister to England, where he staid five months only.

They have no society in Paris like that in London; or if they have I do not know it. But I know no one better able to introduce Mr. Melville to men of education & position, than M. de Beaumont.

In sending these letters to Mr. Melville, I must ask you to request him, not to mention that he has received any letters of introduction from me. I am so often applied to & so often have to refuse, that it is quite desirable to me, when I do it, that it should not be known.

Wishing Mr. Melville an agreeable visit to Europe & the full enjoyment of his reputation, I remain, My dear Sir, as ever, with the highest respect, faithfully Yours —   — Draft — MHS–E

The letter to M. de Beaumont is as follows:

My dear M. de Beaumont,

This note will be handed to You by my ingenious countryman Mr. Herman Melville the author of "Typee," "Omoo" & another work of the same class, which I have not yet read: — publications which have had great success both in Europe & America; & rank with the most original & *piquant* of the day. They have, if I mistake not lately been made the subject of an elaborate criticism in the *Revue des deux Mondes*.

Mr. Herman Melville is about to visit London & Paris. I have not the advantage of knowing him personally. We live as far apart as Paris & Lyons, & you know we have no Paris or London in the U. States, which for some portion at least of the year, gathers all into its net. But I have been requested by my much honored friend, the Chief Justice of the Commonwealth, the father-in-law of Mr. Melville, — to procure for him the advantage of making the Acquaintance of some of those persons in Europe, whom an intellectual American is most desirous of knowing. You must not allow yourself to wonder that I have thought of you; and I beg to assure you that I shall regard any kindness shown to Mr. Melville as doubly shown to myself.

Desiring my kindest remembrance to all the family of Lafayette I remain, My dear M. de Beaumont, ever sincerely yours. — Draft — MHS–E

Friday, December 7

chivalry. Lord Byron's *Childe Harold's Pilgrimage*, Canto III.

[140]

"crane."

In 1830 the original plan was resumed. In 1842 the good work commenced with the thorough repair of the portion of the ch. then in existence. . . . The 2 principal towers, according to the original designs, were to have been raised to the height of 500 ft. The crane employed to lift the stone to the top of the tower stood until 1868. — *Murray's Hand-book Rhine & North Germany* (1877).

amounted to some $2. Alternative reading, "amounted to [the] sum [of] $2."

Three Kings of Cologne.

The *Three Kings of Cologne,* or Magi, who came from the East with gifts for the Infant Saviour. . . . The skulls of the three kings, inscribed with their names — *Gaspar, Melchior,* and *Balthazer* — written in rubies, are exhibited to view through an opening in the shrine, crowned with diadems (a ghastly contrast) which were of gold, and studded with real jewels, but are now only silver gilt. — *Murray's Handbook Rhine & North Germany* (1877).

original picture turned round.

The picture is not one of Rubens's best; the subject is disagreeable, and many travellers think it scarce worth while to pay the fee (1½ mk. for 1 to 3 persons) to the sacristan to turn the picture round, and display the original at the back of the copy. — *Murray's Hand-Book, Rhine & North Germany* (1877).

Girls. Alternative readings, "buying cigar of pretty cigar girl," or "buying cigar[s] of pretty cigar girls."

Quaint. Alternative reading, *great.*

Ehrenbreitstein. Ehrbrincedstein is a characteristic sample of Melville's spelling of foreign proper names — probably the worst.

Metternich. Prince Clemens Wenzel Lothar Metternich-Winneberg (1773–1859), Austrian statesman and diplomatist, the last great representative of the old diplomacy. With the revolutions of 1848, he had fled Vienna and arrived in England where he was to live in complete retirement at Brighton and London. And he had just left England in October, 1849, ready to live out ten more years unshaken in his enmity to greater freedom for the people.

"*Governor of Coney Island.*" A nickname among his sporting friends for "the popular Guilbert Davis," one of the residents of Bond Street, New York, which in the forties "was almost exclusively a social center." — *Reminiscences of Richard Lathers,* edited by Alvan F. Sanborn (New York, 1907).

In 1846 Richard Lathers had married one of the Thurstons of 7 Bond Street, a sister of Allan Melville's wife.

> I have been to-day for the twentieth time, to the Garden of Plants. . . . I went into the cabinets. . . . There I saw . . . some thousands of birds standing on one leg, and looking very sharply at one another, with glass eyes; among them looking equally sharp, a representative of the American Eagle — the "Governor of Coney Island," who says "I am the Governor of Coney Island — probably as well known, as any man in New-York. I've been to Europe fifteen times, Spain five, Russia six, England nine times, and traveled more in France, than any man, that ever lived. Poh! these things are nothing; in Rome they are twice as fine; I have just come from there; I am a wine merchant, that's why I travel so much." — Augustus Kinsley Gardner, M.D., *Old Wine In New Bottles, or Spare Hours of a Student in Paris* (1848).

storied. Melville's lettering of this word is so perfectly "writ in water" that only an authoritative statement by the Geographical Institute, Harvard University, has persuaded the eye from reading *slow.* "Rivers you ask about are *moderately* fast at that point, and therefore you could not properly call them 'slow.'" He had already used the expression, "the storied Rhine" the day before.

Thursday, December 13

Anastasius. Melville made good his loss by purchasing a British edition of *Anastasius,* probably paying more than the four-franc French edition — another instance of the evil workings of the publishing business, because of the lack of an international copyright agreement. The Copyright Bill of 1842 rendered illegal the imports of foreign reprints of British books into England or her Colonies.

So far so good — but not far enough.

Anastasius is interesting for its suggestiveness in relation to the writing of *Moby Dick*, begun on his return from this trip. Anastasius and Anagnosti (another Greek) are both in the bagnio (prison) in Constantinople. Anagnosti speaks,

"Let us become brothers; let religion sanctify our intimacy, so as to divert it of its dangers;" — and upon this he proposed to me the solemn ceremony, which, in our church, unites two friends of either sex in the face of the altar by solemn vows, gives them the endearing appellation of brothers or sisters, and imposes upon them the sacred obligation to stand by each other in life and in death.

— *Anastasius or Memoirs of a Greek*, written at the close of the eighteenth century, by Thomas Hope, Esq., vol I (Paris, 1831), p. 100.

Compare *Moby Dick*, "A Bosom Friend."

He seemed to take to me quite as naturally and unbiddenly as I to him; and when our smoke was over, he pressed his forehead against mine, clasped me round the waist, and said that henceforth we were married; meaning, in his country's phrase, that we were bosom friends; he would gladly die for me, if need should be.

Anastasius, seeking conversion to the Mohammedan faith, is taught by

a personage who, at the end of the Ramadan, looked like a walking spectre, and the very last time of this fast absolutely doubled its length, only for having snuffed up with pleasure, before the hours of abstinence were over, the fumes of a kiebab on its passage out of a cook-shop. — *Anastasius*, p. 136.

See also note to Ramadan, p. 419

"*Ramadan:* or Ramazan: the month during which the Mohammedans fast all day, and feast all night. While the sun remains above the horizon they dare not even refresh themselves with a drop of water or a whiff of tobacco," and compare *Moby Dick*, "The Ramadan of Queequeg."

Duke of Rutland. After Edward Everett's residence in England as U. S. Ambassador from 1841 to 1845, he corresponded occasionally with some of his English friends. Among these was the Duke of Rutland. A letter of his from Cambridge, May 22, 1848, accompanying his "Eulogy on President Adams" expresses "the hope that all the family at Belvoir are in the continued enjoyment of good health," and refers to "the friendly reception & flattering hospitalities which were there extended to us." . . . "Kindest remembrances" are sent by him and his wife and daughter "to Lady Manners & to Lord Granby & the other members of the family at Belvoir, with whom we have the honor to be acquainted." This is another link, perhaps, between Judge Shaw's son-in-law and the Belvoir household, further explaining his invitation.

Everett's use throughout the letter of a plentiful sprinkling of titles — "My Lord Duke" and "Your Grace" — contrasts curiously with Melville's "Mr. Rutland — The Duke of Rutland, I mean" — an early expression of the characteristic conflict in him of aristocratic leanings and democratic urgings. Had he gone to Belvoir Castle, he would have used the proper forms of address, however he might have responded on purely human grounds. — Draft — MHS—E

Belvoir Castle. Listed as one of the twenty-four *Historic Houses of the United Kingdom*, published by Cassell & Company, Limited (London, Paris, & Melbourne, 1892), and described by Charles Edwardes.

The Lord John Manners whom Melville mentions was the Duke in 1892. He allowed "the public to see nearly everything of interest in the Castle." As Mr. Edwardes says "In none of the stately homes of England, indeed, is there less restriction than here." And this was well, for Belvoir's picture gallery was rich in masters — Melville would have seen some of his favorites — Murillo, Claude, Poussin, and the Dutch school. And he would have admired Belvoir's commanding site on a hill overlooking beautiful country. Of its architecture he might well have been critical, since much of the older part was destroyed in a great fire in 1816, and the later additions are described as "Strawberry Hill Gothic." It was probably used as a fortress before the time of William the Conqueror; but it is known that "the foundation of *modern* (the italics are mine) Belvoir dates from the eleventh century."

Powell Papers. Melville must have received, with his letters from home, copies of the *New York Herald* for November 23, 26, and 28, containing the account of a "Frightful Row among the Literati — A Splendid Prospect for Blackwell's Island." It concerns the publication by L. Gaylord Clarke, editor of "The Knickerbocker," of a recriminating letter from Charles Dickens about "one Thomas Powell" (whose book, *The Living Authors of England*, the Appletons had just published) branding him as a forger and a thief, an attempted suicide, a lunatic, and an ingrate. Since Melville knew Powell, he must have been keenly interested in the mixed story of hurt pride, false literary values, financial panic and losses, that the editors of the *Herald* tried to sift in its issue of November 26.

A copy of the first American edition of *Mardi*, inscribed by Melville, "Thomas Powell from Herman Melville, New York, June, 1849," is in the collection of Mr. Bradley Martin, by whose permission I quote the inscription.

Friday, December 14

Paletot. "A kind of loose outer garment or coat for men." Did Melville purchase this at Elias Moses & Son, 83–86 Aldgate Street, a clothing establishment he noticed earlier on one of his walks, and probably recommended to him by one of his New York friends?

green coat.

They were sitting together at a music-hall, — half music-hall, half theatre, which pleasantly combined the allurements of the gin-palace, the theatre, and the ball-room, trenching hard on those of other places. Sir Felix was smoking, dressed, as he himself called it, "incognito," with a Tom-and-Jerry hat, and a blue silk cravat, and a *green coat* [italics mine]. Ruby thought it was charming. Felix entertained an idea that were his West End friends to see him in this attire they would not know him. Anthony Trollope, *The Way We Live Now.*

Mr. Foster. John Foster (1812–1876), literary editor of the *Examiner* 1847–55), and author of *The Life of Goldsmith.*

Mrs. Daniel. In a postscript to a letter of Melville's to Evert Duyckinck, London, December 14, 1849, he writes,

> I this morning did myself the pleasure of calling on Mrs. Daniel for the first. I saw her, & also two very attractive young ladies. Had you seen these young ladies, you would have never told Mrs. Duyckinck of it. You must on no account tell Mrs. Welford of this; for those nymphs were her sisters. — NYPL–D

Duyckinck. For its obvious interest at this point and its intrinsic interest at any point to a Melville reader, I quote from this letter of December fourteenth at length.

> Yesterday being at Mr Bentley's I enquired for his copies of the last "Literary Worlds" — but they had been sent on to Brighton — so I did not see your say about the book Redburn, which to my surprise (somewhat) seems to have been favorably received. I am glad of it — for it puts money into an empty purse. But I hope I shall never write such a book again — tho' when a poor devil writes with duns all round him, & looking over the back of his chair — & perching on his pen & diving in his inkstand — like the devils about St. Anthony — what can you expect of that poor devil? — What but a beggarly "Redburn!" And when he attempts anything higher — God help him & save him! for it is not with a hollow purse as with a hollow balloon — for a hollow purse makes the poet *sink* — witness "Mardi." But we that write & print have all our books predestinated — & for me, I shall write such things as the Great Publisher of Mankind ordained ages before he published "The World" — this planet, I mean — not the Literary Globe. — What a madness & anguish it is, that an author can never — under no conceivable circumstances — be at all frank with his readers. — Could I, for one, be frank with them — how would they cease their railing — those at least who have railed. — In a little notice of "The Oregon Trail" I once said something "critical" about another's [sic] man's book — I shall never do it again. Hereafter I shall no more stab at a book (in print, I mean) than I would stab at a man. — I am but a poor mortal, & admit that I learn by experience & not by divine intuitions. Had I not written & published "Mardi," in all likelihood, I would not be as wise as I am now, or may be.

For that thing was stabbed *at* (I do not say *through*) — & therefore, I am the wiser for it. — NYPL–D

Given in full in Willard Thorp, *Herman Melville.*

Mr. & Mrs. Charles Kean. Charles John Kean (1811?–1868) son of Edmund Kean, actor in many varied parts from early childhood, "careful and conscientious, but scarcely an inspired actor." His greatest part was Louis XI, "Hogarthian" in character. He is best known for his "spectacular revivals" of old classics. — DNB

Ellen Tree Kean (1805–1880), actress before and after her marriage with Charles John Kean. "Of imagination in its highest sense she was deficient, but she had genuine humor and provocative mirth." Westland Marston declares that "in sympathetic emotion, as distinguished from stern and turbulent passion, no feminine artist of her time surpassed her." — DNB

Melville saw her at a time when she was said to be "handsome and intellectual."

Wallack. James William. For biographical note see p. 123.

Jerrold. Douglas William Jerrold (1803–1857), man of letters, author of many plays & magazine articles. His greatest success was "Mrs. Caudle's Curtain Lectures," republished first from *Punch* in 1846.

It is interesting to note that Douglas Jerrold reviewed *Typee* in his Shilling Magazine, London, April 1846. He called the book "one of the most captivating we have ever read. . . . Although, with little pretension to author-craft, there is life and truth in the descriptions, and a freshness in the style of the narrative."

Saturday, December 15

Knight's London. Popular illustrated guidebook by Charles Knight. Melville met Knight at his last London dinner, December 23.

pesky. Melville is quoting a favorite word of his wife's, which I often heard her use.

Lieutenant Wise's book. Henry Augustus Wise (1819–1869), naval officer and author. The book spoken of is *Los Gringos, or an Inside View of Mexico and California, with Wanderings in Peru, Chili, and Polynesia,* published in London in 1849. He went to Tahiti a little after the publication of *Omoo.*

The Reverend H. Melvill. Henry Melvill (1798–1871), at this time Principal of Haileybury College, and for many years the most popular preacher in London, though the style of his rhetoric is said to have appealed more to the literary than the spiritual sense.

Possibly Melville was influenced by more than the name and reputation of the Reverend H. Melvill in his desire to hear him. The *London Times* of Saturday, December 15, carried a long account of the East India Company's Colleges at Haileybury and Addiscombe that existed to train English youth for civil and military service in India. The following passage gives an idea of Haileybury's aims.

> At the College at Haileybury, about 90 gentlemen, between the ages of 17 and 20, are instructed in the oriental languages, in the principles of morals, law, logic, and jurisprudence, and are fitted for the high requirements of the civil service in India. . . . The training he receives is of the character that will best enable him to cope with the subtlety of the Hindu intellect, to track the progress of intrigue in the courts of native princes — and to make himself familiar with all the phases of the Oriental vices of deceit, dissimulation and treachery. To do this efficiently, presupposes no inconsiderable acquaintance with the springs of human action, the laws of the human mind, and the workings of the human heart. These subjects form a portion of the study at Haileybury. . . .

aristocracy. Melville was enough interested to buy a book on the *Aristocracy of England,* of which he made use two years later in the opening book of *Pierre,* "Pierre Just Emerging From His Teens." He seems to have had an inherited interest in his own aristocratic strain, and a natural philosopher's curiosity about the "preservative and beautifying influences of unfluctuating rank, health, and wealth." Note that he wants to know what an aristocratic heritage does to human beings to affect the quality of their feeling and thinking — "what it really and practically is."

resolve it over again. Whoever is tempted to give undue weight to the impulsive forces in Melville, should make a note of this passage. He faced graver situations and made more difficult decisions later in life.

Article upon the "Sunday School Union." This was a letter addressed to "Mr. Editor" and signed "Publicola" in *The Weekly Dispatch,* December 16, 1849, "On Sunday-School Education."

The writer expresses hope for a real movement to consider the subject of National Education and launches into a lengthy and detailed criticism of a system of proselytism that passed for "national education," often accompanied by bitter sectarianism and bigotry, in spite of the "thousands" who "have laboured honestly, liberally, distinterestedly, and most usefully, as voluntary teachers in the schools." It was indeed all the "education" millions of children received.

Why was Melville "struck" by this article? Of course he would agree with a diatribe against sectarianism and bigotry. And separate sentences in the last paragraph would have struck a sympathetic note.

> There is a canting custom in this country of connecting revolutionary movement and infidelity. . . . "The mighty political changes through which we have passed [on the continent] within the last twelvemonths have been over-ruled by our adorable Lord to the greatest of blessings, the bestowment of religious liberty." . . . Democracy is not antagonistic to sincere piety; and there may possibly be many of the Lord's people even among the Red Republicans.

The orthodox phraseology would not have disturbed him. He was familiar enough with it. His sister, Augusta, for whom he had a special attachment, was an ardent Sunday School teacher.

Had he, perhaps, a renewed hope that England would evolve a system of national education more successful than that of his own country? In 1837, during his first experience of teaching in

[149]

Pittsfield, he had written his uncle, Peter Gansevoort (see note p. 169):

> My scholars are about thirty in number, of all ages, sizes, ranks, characters, & education; some of them who have attained the ages of eighteen can not do a sum in addition, while others who have travelled through the Arithmatic: but with so great swiftness that they can not recognize objects in the road on a second journey: & are about as ignorant of them as though they had never passed that way before.

Incidentally one cannot help wondering how Melville fared in the teaching of spelling! See below.

His uncle had sent him John C. Taylor's *District School*. He writes

> Intimatly [sic] am I acquainted with the prevalence of those evils which he alledges [sic] to exist in Common-Schools.
>
> Orators may declaim concerning the universally-diffused blessings of education in our Country, and Essayests [sic] may exhaust their magazine of adje[ctives] in extolling our systim [sic] of common school instruction, — but when reduced to practise, [sic] the high and sanguine hopes excited by its imposing appearance in *theory* — are a little dashed, —
>
> My [sic] Taylor has freely pointed out its defects, and has not been deterred from reproving them, by any feelings of delicasy [sic] — If he had, he would have proved a traitor to the great cause, in which he is engaged. — But I have almost usurped the province of the Edinburgh Review. . . .

"Publicola" perhaps brought back sympathetic memories of Mr. Taylor. Melville was in an impressionable state at the time, after a day of painful decision and homesickness.

I am indebted to Mr. F. C. Francis, of the British Museum, for locating and kindly sending me a photostat of the article on the Sunday School Union.

It may be seen in the Melville Collection, HCL.

30 Days. Melville may have been optimistic about thirty days, for December was considered the worst month to make a westbound passage, the voyage sometimes taking as long as forty-eight days. Apparently one of the indoor sports of passengers aboard ship was to lay bets on the length of the voyage. Despite bad weather and the weight of its cargo, the packet liner usually

[150]

sailed to the utmost limit of its rigging, often at the risk of both sails and sailors.

I am indebted to Ruth Whitman for this note.

Monday, December 17

Powell. Evidently the Thomas Powell of the "Powell papers."

The Independence.

New York. — The only regular Line of Packets between London and New York sail on the 6th, 13th, 21st and 28th of every month, calling at Portsmouth (Messrs. Garrath and Gibbon agents there). The arrangements for cabin, intermediate, and steerage passengers are complete in every respect. The ship now loading is the Independence, 1000 tons, A. T. Fletcher, Commander: lying in the London Docks; to sail December 21st. For freight or passage apply to Messrs. Baring, Brothers and Co., 8 Bishopsgate-street within; or Phillips Shaw, and Lowther, 2 Royal Exchange building. — The London Times, December 17, 1849.

his cousin. Robert Francis Cooke, 4 Elm Court Temple.

Rembrandt's Jew. It is interesting to speculate on which of the two Rembrandt portraits of Jews that were in the National Gallery in 1847 (and so presumably in 1849) Melville would have been most interested in. Of the Rabbi the 1847 catalogue says —

A first-rate specimen. As like nature as Van Dyck: the light is soft and broad, the shadows unobtrusive and transparent. It is beautifully touched in, and the features of the sharp-looking old man have an intelligence and delicacy more in comparison with the fine Van Dyck than Rembrandt's Merchant in this collection. The beard and eyes are most masterly. The only imperfection is a touch of light on the nostril too sudden for sober keeping.

The portrait of the Merchant is described also in glowing terms:

A very powerful representation of an unpolished wealthy individual. The rough features, the grizzly beard, and warm fur cap; the substantial drapery and somewhat clouded splendour of effect are worked to a good purpose with a masterly hand. It is characteristic, picturesque, and possesses a sort of savage dignity.

It remains for some student of Melville's appreciation of art to decide which portrait, at this stage of his development is most

[151]

likely to have been his choice. From black and white reproductions only, the Rabbi would seem to take first place.

The saints of Taddeo Gaddi. Two panels, the descriptions of which may throw light on Melville's interest.

No. 215 — With that simplicity and genuine expression that prevailed before powerful execution and forced effects prevailed, this picture is indeed devoid of perspective and of strength of light and shade and consequently, flat. The colouring is silvery and aerial, the draperies are well arranged and pliant and the lights are true and unbroken.

No. 216 — This picture is very similar to the preceding. The faces are rather more severe of character, the hair and beards remind us somewhat of similar parts in the recently discovered works of art of Ancient Assyria: the general aspect of this piece is rather darker than its companion picture Panel. — *Catalogue of Paintings National Gallery* (1847).

Tuesday, December 18

Revolutionary narrative of the beggar. Israel Potter: His Fifty Years of Exile, published in New York and London in 1855, after having appeared serially in *Putnam's Monthly Magazine*, July 1854 to March 1855.

Chatterton. The Poetical Works of Thomas Chatterton, ed. by Charles B. Wilcox, for W. P. Grant (Cambridge, 1842), in two volumes. Volume I is inscribed in ink, "Herman Melville, London, Dec. 19, 1849." Under this, in pencil, *"Bought at a dirty stall there, and got it bound near by."*

Volume II is inscribed, first, "Herman Melville, London, Dec. 19th, 1849." Under this, a later inscription reads, *"To my Brother John C. Hoadley, Pittsfield, Jan. 6th, 1854. Presented in earnest token of my disclaimer as to the criticism of the word 'friend' used in the fly-leaf of the 'Whale'."*

These volumes of Chatterton contain numerous interesting markings in Melville's usual manner, and others of doubtful origin.

In Volume I is Melville's comment on a statement by the editor, who writes, "It is not unlikely that, had he lived, we might have had another Midsummer Night's Dream; and though

Shakespeare must ever remain unapproachable, still we should have read the rich and exquisite faery poetry of his brother dreamer with delight." After this Melville noted in pencil: *Cant. No man "must ever remain unapproachable."*

In Volume II among other marked passages, is the following of Chatterton's in *Happiness*.

> *Priestcraft! thou universal blind of all,*
> *Thou idol, at whose feet all nations fall;*
> *Father of misery, origin of sin,*
> *Whose first existence did with fear begin;*
> *Still sparing deal thy seeming blessings out.*

— NYPL–B

The beautiful three-volume edition of *The Whale* [or *Moby Dick*] (London, 1851), mentioned in the volume of Chatterton, given to his brother-in-law, John C. Hoadley, in 1854, is extraordinarily interesting for what follows. On the title page is this inscription: *"John C. Hoadley from his friend Herman Melville, Pittsfield Jan. 6th, 1853. "If my good brother John take exception to the use of the word 'friend' here, thinking there is a 'nearer' word; I beg him to remember that saying in the Good Book, which hints there is a 'friend' that sticketh 'closer' than a 'brother'."*

On the right side of front end paper, in Melville's hand, are these words, " *'All life,' says Oken 'is from the sea; none from the continent. Man also is a child of the warm and shallow part of the sea in the neighborhood of the land.'* " — NYPL–B

Guzman. The Rogue or The Life of Guzman de Alfarache. Written in Spanish by Matheo Aleman. Servant to his Catholic Majestie and borne in Sevill.

Literary World's review of Redburn. This long review of "Mr. Melville's *Redburn*" opens and ends with critical comments, but the bulk of it consists of sample passages from the book itself. I give the first paragraph and the last two.

In our last number we called Mr. Melville the De Foe of the Ocean. It is an honorable distinction, to which we think he is fairly entitled by the life-like portraiture of his characters at sea, the strong relishing style in which his observations are conveyed, the fidelity to nature, and, in the combination of all these, the thorough impression and conviction of reality. The book belongs to the great school of nature. It has no verbosity, no artificiality, no languor; the style is always exactly filled by the thought and material. It has the lights and shades, the mirth and melancholy, the humor and tears of real life. . . .

A book of incident and detail cannot be described in an article, but we have suggested to the reader the main outlines of *Redburn*.

In the filling up there is simplicity, an ease, which may win the attention of a child, and there is reflection which may stir the profoundest depths of manhood. The talk of the sailors is plain, direct, straightforward; where imagery is employed the figure being vivid and the sense unmistakable. This sailor's use of language, the most in the shortest compass, may be the literary school which has rescued Herman Melville from the dull verbosity of many of his contemporaries. If some of our writers were compelled to utter a few words occasionally through the breathings of a gale of wind it might benefit their style. There is also much sound judgment united with good feeling in *Redburn* — a knowledge of sailor's life unobtrusively conveyed through a narrative which has the force of a life current from the writer's own heart. — *The Literary World: A Journal of American and Foreign Literature, Science, and Art*, Evert A. & George L. Duyckinck, editors (New York, November 17, 1849).

Wednesday, December 19

pleasant time. In a letter to Mr. Bentley, written June 27, 1850 Melville recalled this pleasant occasion:

. . . . I had a prosperous passage across the water last winter; & embarking from Portsmouth on Christmas morning, carried the savor of the plum-puddings & roast turkey all the way across the Atlantic. But tho' we had a good passage, yet, the little mail of letters with which you supplied me (& by reading the superscriptions of which, I whiled away part of the voyage) hardly arrived in time to beat Her Majesty's Mail by the Cunard Steamer.

I have not forgotten the very agreeable evening I spent in New Brighton Street last winter. Pray, remember me to Mr. Bell & Alfred Crowquill, when you see them.

With compliments to Mrs. Bentley & Miss Bentley, Believe Me
Very Truly Yours
H. Melville

[154]

In the collection of Mr. Bradley Martin of New York, with whose kind permission I quote from it.

Mr. Bell. Undoubtedly Robert Bell (1800–1867), journalist, miscellaneous writer, and friend of Thackeray. He will chiefly be remembered by his annotated edition of the English poets in twenty-four volumes, 1854–57. He died before the work was finished.

Melville acquired Chaucer in this edition at some later period. A correspondence in the autumn of 1876 with his cousin, Catharine Gansevoort Lansing, shows that he has been trying to procure for her husband a set "of the old poet who didn't know how to spell, as Artemus Ward said." By October 12 he had found one in a Nassau Street bookshop. He wrote, "The Chaucer is in eight volumes. – good print – same edition as mine – Bell's – but it is perfect." He had evidently come to respect the man who in 1849 was merely "connected with Literature in some way or other." – NYPL–GL – Reproduced in *The Family Correspondence of Herman Melville* by Victor Hugo Paltsits (1929).

("Alfred Crowquill"). Alfred Henry Forrester (1804–1872), best known as Alfred Crowquill, prolific comic writer, illustrator, caricaturist, and writer of tales for children.

Mr. Moore. Mr. Moore of Moore and Illingworth, surgeons, 1 Arlington Street. – *London Directory (1847)*.

No. 1 St. James Place. Probably the Railway Coffee House. – *London Directory (1847)*.

Thursday, December 20

glorious time.

It was the very perfection of quiet absorption of good living, good drinking, good feeling, and good talk. We were a band of brothers. Comfort – fraternal, household comfort, was the grand trait of the affair. Also you could plainly see that these easy-hearted men had no wives or children to give an anxious thought. Almost all of them were travellers, too; for bachelors alone can travel freely, and without any twinges of their con-

[155]

sciences touching desertion of the fireside. — *The Paradise of Bachelors and the Tartarus of Maids. Billy Budd and Other Prose Pieces.*

Cunningham. Peter Cunningham (1816–1869). His *Handbook for London,* just published in June, is distinguished by the literary merit of its quotations from eminent authors. He was treasurer of the Shakespeare Society, editor of the works of Drummond of Hawthornden in 1833, and of *Walpole's Letters* in 1857. Melville owned a copy of Walpole's Letters.

Comical. This word had been deleted.

Woodfall. George Woodfall (1767–1844), who also reprinted *Hakluyt's Voyages.*

Leslie. Charles Robert Leslie (1794–1859), painter and illustrator, was born in London of American parents, spent most of his childhood in Philadelphia, but returned to London as a youth to study art. One of his portraits of Walter Scott is in Harvard's Houghton Library. Melville met Leslie December 23.

5th story.

The apartment was well up toward heaven, I know not how many strange old stairs I climbed to get to it. But a good dinner, with famous company, should be well earned. No doubt our host had his dining-room so high with a view to secure the prior exercise necessary to the due relishing and digesting of it. — *The Paradise of Bachelors and the Tartarus of Maids. Billy Budd and Other Prose Pieces.*

Paradise of Bachelors. *The Paradise of Bachelors and the Tartarus of Maids* first appeared in Harper's New Monthly Magazine, April, 1854. Also in *Billy Budd and Other Prose Pieces.*

Mr. Rogers. Samuel Rogers (1763–1855).

Rogers is silent, — and, it is said, severe. When he does talk, he talks well; and on all subjects of taste, his delicacy of expression is pure as his poetry. If you enter his house — his drawing-room — his library — you of yourself say, this is not the dwelling of a common mind. There is not a gem, a coin, a book thrown aside on his chimney-piece, his sofa, his table, that does not bespeak an almost fastidious elegance in the possessor. But this very delicacy must be the misery of his existence. Oh the jarrings his disposition

must have encountered through life! — *Letters and Journals of Lord Byron:
with notices of his life*, by Thomas Moore (London), 1833, v. I, p. 550.

Mr. Cooke. Robert Francis Cooke, 4 Elm Court, Temple. His
visiting card was kept by Melville's wife, and is now in my possession. On the back of it is written "Mr. John Tenniel" (in an un-
known hand) under which Melville has written "Dryden St. Cec.,"
and in the upper right hand corner on the same side, "Cope &
Herbert Cordelia," references which are explained in the text.

> But all Templars are not known to universal fame; though if the having
> warm hearts and warmer welcomes, full minds and fuller cellars, and giving
> good advice and glorious dinners, spiced with rare divertisements of fun
> and fancy, merit immortal mention, set down, ye muses, the names of
> R. F. C. and his imperial brother. — *The Paradise of Bachelors and the
> Tartarus of Maids*, in *Billy Budd and Other Prose Pieces*.

Cooke's brother. William Henry Cooke, barrister of 4 Elm
Court, Temple.

Reform Club House. The Reform Club was founded by the
Liberal membership of the two Houses of Parliament, about the
time the Reform Bill was canvassed and carried, 1830–32. It was
famous for its French cuisine, as well as for its imposing architecture.

Mr. John Tenniel. John Tenniel (1820–1914), probably more
famous for his illustrations of Lewis Carroll's *Alice In Wonder-
land* and *Alice Through The Looking-Glass* than for any more
ambitious work he ever undertook.

"Herbert." John Rogers Herbert (1810–1890), painter of portraits and historical subjects. In view of the deep influence King
Lear had on Melville, his special mention of this painting is interesting. See Charles Olson, *Lear and Moby Dick*, in *Twice A
Year* (Fall-Winter, 1938).

Cope. Charles West Cope (1811–1890).

Too late. He was too late to see the main performance, *The
Mountaineers*; but he evidently saw a rehearsal of the pantomime (an afterpiece which opened December 26) entitled *The*

Moon Queen and King Night; or *Harlequin Twilight*, to which "Crowquill" had invited him when they met at Mr. Bentley's dinner the night before.

Mr. Miller. John Miller of Tavistock Street, Covent Garden, Despatch Agent of the U. S. Legation.

Erectheum Club. Celebrated for good dinners. Once the town depot of Wedgewood's famous "ware."

Mr. Cleaves. Presumably Mr. Cleaves of 9 King's Bench Walk, Temple, listed at the end of the Journal.

The Mall. Short for the street, Pall Mall, vulgarly called Pell Mell from the French game *Pale Maëlle*, introduced and played here in the reign of Charles I. It became one of London's most "clubable" streets.

chimney-place. In the autumn of the following year Melville bought an old farmhouse in Pittsfield, Massachusetts, which had a massive central chimney and great *chimney-place* in what was once a "sublime kitchen." He named the house "Arrowhead." See *I and My Chimney,* published in *Putnam's Monthly Magazine* for March 1856: also in *Billy Budd and other Prose Pieces* (1924). See also *Herman Melville's "I And My Chimney"* by Merton M. Sealts in *American Literature,* Vol. 13, No. 2, (May, 1941), for an interpretation of the significance of the great chimney to Melville.

Cottenham. Charles Christopher Pepys, Earl Cottenham (1781–1851), was Lord High Chancellor 1836–1841 and 1846–1850. — *Encyclopedia of Chronology, Historical and Biographical,* edited by Woodward and Cates (1872).

Master of the Rolls. In 1849, Henry Bickersteth Lord Langdale (1783–1851), a liberal, independent, reforming judge.

"Master of the Rolls, an equity judge, derives his title from having the custody of all charters, patents, commissions, deeds,

[158]

and recognizances, entered upon rolls of parchment." — *Haydn's Dictionary of Dates* (1906).

Mr. Foster. John Foster (1812–1876), literary editor of the "Examiner" 1847–55, and author of *The Life of Goldsmith.*

Saturday, December 22

Mr. Davidson. A letter written to George Duyckinck by Davidson, Aldine Chambers, 13 Paternoster Row, London, December 24, 1849, gives an account of his meetings with Melville. Other excerpts from it, besides his description of the meetings, throw a pleasant light on a "good fellow." He writes —

I am very much obliged to you for an introduction to Hermann Melville Esq. — we passed two evenings together which considering the dinners and breakfasts he had to do here was wonderfully fortunate — I suppose I was not exactly the kind of man to bring him out — still we had a pleasant brace of evenings. one at the Mitre Tavern, Fleet St. where we sat right opposite to Dr. Johnson's corner in which there was a manner [sic] eating stewed mutton for dinner with his pint of 'af n'af at six o'clock in the evening — Doing our well done 'steaks we adjourned (by omnibus) to the "Blue Posts" Cork Street, Burlington Arcade and had a pint of real punch at the very table you, and Boot *lare* (Butler) and I blue posted a dinner once upon a time. The second evening was at the Blue Posts for the entire thing commencing on beefsteaks and ending on empty pitchers, or in other words from six till half past eleven — damages — paid by Melville 14/2 sterling. We asked each other in the small smoking room — which we had to ourselves with a fine fire — Where, — said he — Where — said I — where said we bothe together — where in America can you find such a place to dine and punch as this?

We talked of you, of several New Yorkers, some books, his affairs, and a tour he had made. He has succeeded most admirably in his business here and leaves tomorrow in the Independence. He went up the Rhine, to Paris and otherwise did a small tour.

This puts me in mind of my tour to those castellated diggings. You say the last you heard of me I was on the Rhine and that you did not [doubt] but that I was in it. Somewhere in August I struck the Rhine at Cologne after perambulating about that kingdom of four million people — Belgium and arriving dirty and tired and hungry and recollecting from Murray that I should show some enthusiasm I went to the river, entered its bosom and dived, kicked, and swam about until my increased hunger made me so

[159]

buoyant I verily believe I could have floated all the way to Holland. My dinner at Cologne that day — did astonish the natives. . . .

Little village near Bonn — Poppelsdorf — walked behind a funeral with my hat in hand all up one street an amateur to get out into the country. . . . Walked to Konigswinter — and at nine o'clock in the morning I had scaled the Drachenfels — staid there two hours. Up come two German students — we converse in French one a lawyer, t'other a doctor — both of the right sort. We sang — "Am Rhein" "Marsellaise" God Save the Queen and Yankee Doodle — told them I was a Doodle — then on top where they sell wine — drank two bottles price 16 cents and boy offered us a pot of red paint to put our names on immortality (the ruins) — very respectfully declined. We descended and went over to Rolandseck another ruin w[h]ere we kissed each others moustaches and parted. . . .

"I toiled up [Marienburg] and had a heavenly view. Besides I noticed in the visitors book the following

Albert Smith   Author   England

very beautifully done — isn't it. . . .

In short I went to Mayence, Franckfurt (to see Daneker's Ariadne, the cemetery, and Goethe's house, and "Stifts beer house" where they cant a barrel up on the counter and sell away) and all over the Rhine, spending over three weeks walking about enjoying myself with one clean shirt, one collar, one pair of socks, and everything I could see — Selah — NYPL–D

Sunday, December 23

Charles Knight, (1791–1873), author and publisher, son of Charles Knight, the bookseller, in whose shop, as apprentice, he received most of his education. He was interested throughout his life in bringing all kinds of knowledge within reach of the poorest. Always a liberal, he was one of the earliest members of the Reform Club. *Half-Hours with the Best Authors* begun in 1847 · is but one of the numerous publications — books, articles, magazines, which he wrote, or which he inspired.

His public spirit is well expressed in his plan for a *Popular History of England* (1862) — "to trace through our annals the essential connection between our political history and our social," to enable the people "to learn their own history — how they have grown out of slavery, out of feudal wrong, out of regal despotism — into constitutional liberty, and the position of the greatest estate in the realm." —DNB

Ford. Richard Ford (1796–1858), Spanish traveler, collector of Spanish art, and author of *The Handbook for Travellers in Spain,* of which it has been said that "so great a literary achievement had never before been performed under so humble a title." — DNB

Barry Cornwall. Bryan Waller Procter (1787–1874), editor, writer of graceful songs, amiable friend of genius, biographer of Charles Lamb.

A plain, middle-sized, rather smallish, English-looking gentleman, elderly (sixty or thereabouts) with short white hair. Particularly quiet in his manners; he talks in a somewhat feeble tone and emphasis, not at all energetic, scarcely distinct. An American of the same intellectual calibre would have more token of it in his manner and personal appearance, and would have a more refined aspect; his head, however, has a good outline, and would look well in marble; but the English complexion takes greatly from its chasteness and dignity. I liked him very well; he talked unaffectedly, showing an author's regard to his reputation, and evidently was pleased to hear of his American celebrity. Nothing remains on my mind of what he said, except that in his younger days, he was a scientific pugilist, and once took a journey to have a sparring encounter with the Game-Chicken. Certainly, no one would have looked for a pugilist in this subdued old gentleman. . . . He is slightly deaf, and this may be the cause of his feeble utterance — owing to his not being able to regulate his voice exactly by his own ear. On the whole, he made a pleasant and kindly, but not a powerful impression on me — as how should he? being a small, though elegant poet, and a man of no passion or emphasized intellect. — Hawthorne's Journal of June 12, 1854, from *The Heart of Hawthorne's Journals,* edited by Newton Arvin (Boston, 1929).

Kinglake. Alexander William Kinglake (1809–1891), author of *Eöthen* at the time Melville met him.

Since the memories of "a pondering man" are deep and lasting, his personal response to Kinglake may be gathered from a marked passage ("The Literary Influence of Academies, p. 68), in his copy of Arnold's *Essays in Criticism* (Boston, 1865), in which he wrote his name, "H. Melville, July 10, '69, N.Y." HCL–M

Arnold has assigned "the Corinthian style" — "the style 'for a good editorial' " to Kinglake's *Invasion of the Crimea.* Melville

[161]

marks the next passage with a single marginal line, adding two extra lines beside the last sentence, and a cross referring to the significant personal note below. He underlines "rapidity without ease" and double underlines "effectiveness without charm."

It has not the warm glow, blithe movement, and soft pliancy of life, as the Attic style has; it has not the over-heavy richness and encumbered gait of the Asiatic style; it has glitter without warmth, rapidity without ease, effectiveness without charm. Its characteristic is, that it has no soul; all it exists for, is to get its ends, to make its points, to damage its adversaries, to be admired, to triumph.

+ The style is (in the case of Mr. Kinglake) eminently the man.

Kinglake's approach to writing must have been very different from Melville's.

He had about 1835 made the Eastern tour described afterwards in *Eöthen, or Traces of Travel brought home from the East.* . . . The book . . . . was the result of a third attempt after he had twice failed to satisfy himself, and did not appear until 1844. . . . He has endeavoured, he adds, and he thinks successfully, to exclude from it "all valuable matter derived from the works of others." — DNB

Melville's debt to other writers was high, not only consciously, but in his deepest being.

And as the great Mississippi musters his watery nations: Ohio, with all his leagued streams; Missouri, bringing down in torrents the clans from the highlands; Arkansas, his Tartar rivers from the plain; — so, with all the past pouring in me, I will roll down my billow from afar — *Mardi,* Chapter XV, "Dreams."

See the whole chapter.

Mr. George Atkinson. Probably Mr. George Atkinson, barrister, special pleader, Northern Circuit, 1 Inner Temple Lane. — *London City Directory (1847).*

4th story of a house.

Where else could I go for rest, unless I crawled into my cold and lonely bed far up in the attic of Craven Street, looking down upon the muddy Phlegethon of the Thames. — *The Two Temples — Temple Second, Billy Budd and Other Prose Pieces.*

The manuscript of *The Two Temples* is in the Melville Collection at the Harvard College Library.

Quebec Hotel. Situated "on the west side of Bath Square. The rear of the Hotel overlooked the harbour, and if there had been windows on that side, the view from them would have been extremely interesting."

The Quebec Hotel was closed round about 1860. When long-distance passenger vessels ceased to call at Portsmouth, its trade must have declined. Part of the original building remains and is used as offices by a shipping agent.

I am indebted to the courtesy of Mr. H. Sargeant, City Librarian and Curator, Central Library, Portsmouth, for this information.

Somerby. The *New York Herald* of February 1, 1850 carries the following notice:

Arrived.

Packetship Independence, Fletcher, London and Portsmouth, Dec. 25, with mdse. and 52 passengers. —

Among the "Passengers Arrived" are listed "*H. Melville, of New York; H. G. Somerby, of Newburyport.*"

I am indebted to Mr. Jay Leyda for calling this notice to my attention.

butters. Melville evidently delighted in this ancient word to describe the succulent birds hung with the cutlets.

Butte. *obsolete form of* Bittern. . . . Skaildraik, Herron, Butter, or any
sic kynde of fowlles.

Bittern. *A genus of grallatorial* [wading] *birds* (Botaurus) *nearly allied to the herons, but smaller.* . . . *The patriche, plover, bittorn and heronsewe.* — New English Dictionary (Oxford, 1888).

Mrs. Shaw. Hope Savage Shaw, second wife of Judge Lemuel Shaw and step-mother of Melville's wife. This gift is recorded in a letter from Melville to Mrs. Shaw, Sunday afternoon, N.Y. [early 1850] when he speaks of sending her a "University Bread Trencher" via "Sammie" (his brother-in-law, Samuel.) — HCL–M

See *supra* (p. 70) the bread Trencher he bought on December 15.

[163]

Portsmouth. "Since contrary winds often meant considerable delay in going around from London to the South coast, the London packets would pick up or leave mails and passengers at nearby Portsmouth, the greatest of naval bases." — R. G. Albion, *Square-Riggers on Schedule* (Princeton, 1938), p. 34.

Tuesday, December 25

Melville had actually dated this entry Tuesday, December 26, Christmas — no doubt in his confusion over sailing.

"North Corner." "Although the expression may have been current in 1849, there is no record of it," according to Mr. Sargeant, City Librarian and Curator, Central Library, Portsmouth. But that it applies to some part of the "Point," and that its "fame" was confined to the sailor world with which Melville was familiar, seems obvious. In *Redburn* he had already written —

> During the greater part of the watch, the sailors sat on the windlass and told long stories of their adventures by sea and land, and talked about Gibraltar, and Canton, and Valparaiso, and Bombay, just as you and I would about Peck Slip and the Bowery. Every man of them almost was a volume of Voyages and Travels round the World. And what most struck me was, that like books of voyages they often contradicted each other, and would fall into long and violent disputes about who was keeping the Foul Anchor tavern in Portsmouth at such a time; or whether the King of Canton lived or did not live in Persia; or whether the barmaid of a particular house in Hamburg had black eyes or blue eyes; with many other mooted points of that sort. — *Redburn,* Chapter IX.

In *Israel Potter,* published six years later, he was to write —

> The officers being landed [in Portsmouth] some of the crew propose, like merry Englishmen as they are, to hie to a neighboring ale-house, and have a cosy pot or two together. — *Israel Potter,* Chapter III.

Concerning the geography of Portsmouth and the character of the "Point" Mr. Sargeant writes:

> The geography of Portsmouth was quite different one hundred years ago. At that time the borough consisted, in the main, of two adjacent parishes, Portsmouth and Portsea, each separately enclosed by an elaborate system of fortifications. To the west of Portsmouth, outside the walls, was a small peninsular known as "Point." This had for long been the favourite resort of sailors, for whom it had three advantages — it had an extraordinary

[164]

number of brandy shops and beer houses, it was outside the town walls and therefore beyond the jurisdiction of the military Governor, and it was conveniently close to the Harbour.

"Point" was on the way to becoming a respectable place when Melville saw it, but during the 18th and early 19th centuries it was most unsalubrious. (The caricaturist Rowlandson drew a well-known picture of it when it was in its heyday.)

I am indebted to Mr. Sargeant of the Central Library, Portsmouth, for several letters and two maps, evincing remarkable interest in Melville's relation to Portsmouth.

Nelson's ship. Forty years or more later Melville wrote —

Nevertheless, to anybody who can hold the Present at its worth without being inappreciative of the Past, it may be forgiven, if to such an one the solitary old hulk at Portsmouth, Nelson's *Victory,* seems to float there, not alone as the decaying monument of a fame incorruptible, but also as a poetic reproach, softened by its picturesqueness, to the *Monitors* and yet mightier hulk of the European ironclads. — *Billy Budd and Other Prose Pieces* Chapter IV.

Melville's warm, personal admiration for Nelson is given explicit expression throughout this chapter. He heads it "Concerning 'The greatest sailor since the world began' " — Tennyson's phrase. He refers to "the genius of Nelson," and ends with — "the poet but embodies in verse those exaltations of sentiment that a nature like Nelson, the opportunity being given, vitalizes into acts.

Hawthorne links Tennyson and Nelson in an interesting passage in his journal of July 30, 1857, when he had seen Tennyson in an art gallery.

He [Tennyson] is exceedingly nervous, and altogether as un-English as possible; indeed an Englishman of genius usually lacks the national characteristics, and is great abnormally. Even the great sailor, Nelson, was unlike his countrymen in the qualities that constitute him a hero; he was not the perfection of an Englishman, but a creature of another wind — sensitive, nervous, excitable, and really more like a Frenchman. — *The Heart of Hawthorne's Journals,* edited by Newton Arvin (Boston, 1929).

Whether it was because of natural interest in all aspects of Nelson's life, or appreciation of what Lady Hamilton meant to him, or admiration for Romney's many portrayals of her beauty

— or something of all three — Melville had among his engravings
one of Lady Hamilton as a Bacchante. There is no record of his
having seen the original, now in the National Gallery, as it was
in the South Kensington Museum in 1849.

Memoranda of Things on the Voyage

"*Opium Eater.*" One character in De Quincey's *Confessions of
an English Opium Eater*, that "most wondrous book," may have
had a suggestive effect when Melville came to write *Moby Dick*.
Fedallah and his companions seem to owe something of terror
and strangeness to "the Malay."

The following passages are from *Confessions of an English
Opium Eater* (1850).

"One day a Malay knocked at my door — possibly he was on
his way to a sea-port about forty miles distant." He wore
"turban and loose trousers of dingy white," and appeared like "a
tiger-cat" to De Quincey's country-bred maid.

> . . . this Malay . . . fastened afterwards upon my dreams, and brought
> other Malays with him worse than himself, that ran "a-muck" at me, and
> led me into a world of trouble. . . .
>
> The Malay had been a fearful enemy for months. I have been every
> night, through his means, transported into Asiatic scenes. . . . Southern
> Asia, in general, is the seat of awful images and associations. . . . The mere
> antiquity of Asiatic things, of their institutions, histories, modes of faith,
> &c., is so impressive, that to me the vast age of the race and name over-
> powers the sense of youth in the individual. A young Chinese seems to me
> an antedeluvian man renewed.
>
> The cursed crocodile became to me the object of more horror than almost
> all the rest. I was compelled to live with him; and (as was always the case
> almost in my dreams) [sic] for centuries.

Compare these with the following passages from *Moby Dick*:

> The figure that now stood by its bows was tall and swart, with one white
> tooth evilly protruding from its steel-like lips. A rumpled Chinese jacket
> of black cotton funereally invested him, with wide black trowsers of the
> same dark stuff. But strangely crowning this ebonness was a glistening
> white plaited turban, the living hair braided and coiled round and round
> upon his head. Less swart in aspect, the companions of this figure were of

that vivid, tiger-yellow complexion peculiar to some of the aboriginal natives of Manilla; — a race notorious for a certain diabolism of subtlety, and by some honest white mariners supposed to be the paid spies and secret confidential agents on the water of the devil, their lord. . . .

. . . The White Whale now shook the slight cedar as a mildly cruel cat her mouse. With unastonished eyes Fedallah gazed, and crossed his arms; but the tiger-yellow crew were tumbling over each other's heads to gain the uttermost stern.

schoolmaster. The four following notes on school teaching probably reflect feelings resulting from two experiences in Melville's earlier years. In 1837, when he was only eighteen, he had taught a term in a country public school in the Sykes district of Pittsfield, "a remote and secluded part of the town about five miles from the village." Again in 1839, after his trip to Liverpool, he taught school in Greenbush, New York, now East Albany. It is not known why he gave up teaching in Pittsfield; but his second venture in Greenbush ended in the financial collapse of the school during a depression.

Considering that some of his pupils were the same age as Melville in 1837, it is not improbable that he sometimes encountered difficulty in controlling and teaching them.

"could have killed his scholars sometimes."

M. Deybens suggested to me that I should undertake the education of M. de Mably's children. . . . I possessed almost sufficient knowledge for a tutor, and believed I had the necessary qualifications. During the year which I spent at M. de Mably's, I had ample time to undeceive myself. My naturally gentle disposition would have made me well adapted to this profession, had not a violent temper been mingled with it. As long as all went well, and I saw that my trouble and attention, of which I was not sparing, were successful, I was an angel; but when things went wrong, I was a devil. When my pupils did not understand me, I raved like a madman; when they showed signs of insubordination, I could have killed them, which was not the way to make them either learned or well behaved. — *The Confessions of Jean Jacques Rousseau* (Edinburgh, 1904), vol. I, p. 275–276.

"intolerable."

In the forlorn state of his circumstances, he accepted of an offer to be employed as usher, in the school of Market-Bosworth, in Leicestershire. . . .

This employment was very irksome to him in every respect — it was unvaried as the note of the cuckoo; and — he did not know whether it was more disagreeable for him to teach, or the boys to learn, the grammar rules. — James Boswell, *Life of Samuel Johnson. . . .,* 10 vol. (London, 1839), vol. 1, p. 87.

In the house of Sir Wolstan Dixie, the patron of the school, he was treated with what he represented as intolerable harshness.

"The forlorn state" of Johnson when he became an usher, and the dryness of the occupation that he found "intolerable," find an imaginative combination when Melville ascribes the etymology of the whale as follows:

Etymology
(Supplied by a late consumptive usher to a grammar school) The pale Usher — threadbare in coat, heart, body, and brain; I see him now. He was ever dusting his old lexicons and grammars, with a queer handkerchief, mockingly embellished with all the gay flags of all the known nations of the world. He loved to dust his old grammars; it somehow mildly reminded him of his mortality. — *Moby Dick.*

Other allusions.

The truth, however, is, that he was not so well qualified for being a teacher of elements, and a conductor in learning by regular gradations, as men of inferior powers of mind. His own aquisitions had been made by fits and starts, by violent irruptions in the regions of knowledge; and it could not be expected that his impatience would be subdued, and his impetuosity restrained, so as to fit him for a quiet guide to novices. The art of communicating instruction, of whatever kind, is much to be valued; and I have ever thought that those who devote themselves to this employment, and do their duty with diligence and success, are entitled to very high respect from the community, as Johnson himself often maintained. Yet I am of opinion, that the greatest abilities are not only not required for this office, but render a man less fit for it. — Boswell, vol. I, p. 104.

Johnson was not more satisfied with his situation as the master of an academy, than with that of the usher of a school; we need not wonder, therefore, that he did not keep his academy above a year and a half. — Boswell, vol. I, page 105.

Goldsmith had taken lodgings at a farmer's house in order to have full leisure for writing. "He said, he believed the farmer's family thought him an odd character, similar to that in which the *Spectator* appeared to his landlady and her children: he was *The Gentleman*." — Boswell, vol. 3, p. 220.

This attitude of Goldsmith's landlord must have struck Melville as an interesting contrast to that of the Yankee farmer with whom he boarded during his teaching term in Pittsfield, when he also had a little leisure for "occasional writing."

The man with whom I am now domicilated [sic] is a perfect embodiment of the traits of Yankee character, — being shrewd bold and independant [sic], carrying himself with a genuine republican swagger, as hospitable as "mine host" himself, perfectly free in the expression of his sentiments, and would as soon call you a fool or a scoundrel, if he thought so — as, button up his waistcoat.

From a letter to his uncle, Peter Gansevoort, dated Pittsfield, December 31, 1837 — NYPL–GL. Reproduced in *Family Correspondence of Herman Melville*, edited by Victor Hugo Paltsits (New York, The New York Public Library, 1929).

Cannibals. Melville must have glanced at the marginal note on page 17, book I of *Vulgar Errors* — or, as the old folios had it *Pseudodoxia Epidemica or Enquiries into Very many Received Tenents And commonly presumed Truths, by Thomas Browne, Dr. of Physick*. This simply reads, "Eating of Man's flesh," and refers to a passage intended to correct the "vulgar error" which held that Acteon was devoured by his hounds, and Diomedes by his horses ("Anthropophagie of Diomedes his horses" — anthropophagie being a rare derivative of anthroprophogi, man-eating.)

Ben Jonson was writing humorously, if a bit gruesomely, of the real thing. —

Marry, I say, nothing resembling man more than a swine, it follows, nothing can be more nourishing; for indeed (but that it abhors from our nice nature) if we fed one upon another, we would shoot up a great deal faster, and thrive much better; I refer me to your usurous cannibals, or such like: but since it is so contrary, pork, pork, is your only feed. — *Every Man Out Of His Humour*.

[169]

Indian.

Next was Tashtego, an unmixed Indian from Gay Head, the most westerly promontory of Martha's Vineyard, where there still exists the last remnant of a village of red men, which has long supplied the neighboring island of Nantucket with many of her most daring harpooneers. — *Moby Dick.*

Gay Head.

On the farthest tip of Martha's Vineyard [an island off Cape Cod, Massachusetts] across Menemsha Pond, rises a peninsula that ends in cliffs composed of strata of incredibly variegated clays — red, blue, orange, tan, and black — alternating with dazzling white sandy substance. This is Gay Head. — *Massachusetts. A Guide To Its Places And People* (Boston, 1937).

Indian (Gay-Head) Sweetheart flogged. The Christian Indians of Gay Head, converted in the seventeenth century, frowned on perpetuating stories of witches, devils, and violent crimes. Some of these persisted popularly, however, well into the nineteenth century; and though no "sweetheart flogged" came to light in the course of many conversations with the courteous and interested older inhabitants, one of them, Mrs. Anna Hayson (now over seventy) told me that her father would have kept from his children just such a story. Many Gay Head Indian folk tales have been lost in this way. But the most important legends, including those about *a great white whale,* are still alive. Melville could have heard them aboard the *Acushnet* — the *Pequod* of *Moby Dick.* See *The Weather Breeder* by Sylvia Chatfield Bates for imaginative use of Gay Head Indian material.

Fly — wounding our clothes. Ben Jonson describes the duel as follows:

Fastidious Brisk. Good faith, signior, now you speak of a quarrel, I'll acquaint you with a difference that happened between a gallant and myself; Sir Puntarvolo, you know him if I should name him, signior Luculento.
Puntarvolo. Luculento! what inauspicious chance interposed itself to your two loves?
Fast. Faith, sir, the same that sundered Agamemnon and great Thetis' son; but let the cause escape, sir: he sent me a challenge, mixt with some few braves, which I restored, and in fine we met. Now, indeed, sir, I must tell you, he did offer at first very desperately, but without judgement: for, look

you, sir, I cast myself into this figure; now he comes violently on, and withal advancing his rapier to strike, I thought to have took his arm, for he had left his whole body to my election, and I was sure he could not recover his guard. Sir, I mist my purpose in his arm, rashed his doublet-sleeve, ran him close by the left cheek, and through his hair. He again lights me here, — I had on a gold cable hatband, then new come up, which I wore about a murrey French hat I had, — cuts my hatband, and yet it was massy goldsmith's work, cuts my brims, which, by good fortune, being thick embroidered with gold twist and spangles, disappointed the force of the blow: nevertheless, it grazed on my shoulder, takes me away six purls of an Italian cut-work band I wore, cost me three pound in the Exchange but three days before.

Punt. This was a strange encounter.

Fast. Nay, you shall hear, sir: with this we both fell out, and breath'd. Now, upon the second sign of his assault, I betook me to the former manner of my defence; he, on the other side, abandon'd his body to the same danger as before, and follows me still with blows: but I being loth to take the deadly advantage that lay before me on his left side, made a kind of stramazoun, ran him up to the hilts through the doublet, through the shirt, and yet miss'd the skin. He, making a reverse blow, — falls upon my emboss'd girdle, I had thrown off the hangers a little before — strikes off a skirt of a thick-laced satin doublet I had, lined with four taffatas, cuts off two panes embroidered with pearl, rends through the drawings-out of tissue, enters the linings, and skips the flesh.

Carlo Buffone. I wonder he speaks not of his wrought shirt.

Fast. Here, in the opinion of mutual damage, we paused; but, ere I proceed, I must tell you, signior, that, in this last encounter, not having leisure to put off my silver spurs, one of the rowels catch'd hold of the ruffle of my boot, and, being Spanish leather, and subject to tear, over-throws me, rends me two pair of silk stockings, that I put on, being some-what a raw morning, a peach colour and another, and strikes me some half inch deep into the side of the calf: he, seeing the blood come, presently takes horse, and away: I, having bound up my wound with a piece of my wrought shirt —

Car. O! comes it in there?

Fast. Rid after him, and, lighting at the court gate both together, embraced, and march'd hand in hand up into the presence. Was not this business well carried?

Macilente. Well! yes, and by this we can guess what apparel the gentleman wore.

Punt. 'Fore valour, it was a designement begun with much resolution, maintain'd with as much prowess, and ended with more humanity. — *Every Man Out Of His Humour.*

so long as you do it without blushing. Melville remembered the sense of a speech from Ben Jonson's *Every Man Out of His Humour,* if not the exact words.

Sogliardo, "an essential clown," is being prepared by Macilente, an envious "gentleman" for a part which will make him the butt of ridicule.

Maci. Come on, signior, now prepare to court this all-witted lady, most naturally, and like your self.

Sog. Faith, an' you say the word, I'll begin to her in tobacco.

Maci. O. fie on't: no: you shall begin with, *How does my sweet Lady,* or, *Why are you so melancholy, Madam?* though she be very merry, it's all one: be sure to kiss your Hand often enough; pray for her health, and tell her, how, *more than most fair* she is. Screw your Face at' one side thus, and protest; let her fleer, and look a scew, and hide her Teeth with her Fan, when she laughs a Fit, to bring her into more matter, that's nothing: you must talk forward (though it be without sense, so it be without blushing) 'tis most Court-like, and well.

2 copies of "Redburn." A copy of *Redburn* in two volumes is inscribed by Melville, "*Maria G. Melville from Her Affectionate Son, Herman. Pittsfield, January, 1850*" — NYPL–B

This is almost certainly one of the copies he brought with him from London in 1849.

The other copy was probably that given to Dr. Augustus K. Gardner, to whom he seems to have been indebted for the address of Madame Capelle, with whom he lodged in Paris.

The note which accompanied the gift reads:

Monday Morning [Feb. 9 — 1850]

Dear Gardiner — Will you do me the favor to accept the accompanying set of "Redburn" as a slight token of my having remembered you while away. — I lodged with Madame Capelle in Paris & will tell you what I saw in that gay city, when I am so happy as to see you again.

Sincerely yours,

H. Melville

— Essex Institute, Salem

This acknowledges his indebtedness to Dr. Gardner not only for Madame Capelle's address, but for the gift to him of Dr. Gardner's book, *Old Wine in New Bottles* (see Introduction). This volume was in the possession of the late Carroll A. Wilson of New York. I am indebted to Mr. Jay Leyda for calling my attention both to the book and to Melville's note.

Vathek. William Beckford's *Vathek: An Arabian Tale.* The influence of this book may be traced to "L'Envoi," the last poem in *Timoleon*, published by Melville in 1891, where the mountain, Kaf, takes a brief but important place. Under "the terrible Kaf" lay the Subterranean Palace of Fire, the Arabian Hell.

According to a note in the first French edition (Paris, 1787), Kaf was in geographical fact the Caucasus, so enormous that it was thought to encircle the earth. But in Arabian legend it had portentous significance. It was said to have for its base a stone, one grain of which gave the power to perform miracles. This stone was represented as the pivot of the earth, and was like a vast emerald, whose reflected rays gave to the skies their azure. It was thought that when God wanted to cause an earthquake, he commanded this stone to move one of its fibres (which were like nerves) and the neighboring earth trembled, sometimes to great destruction.

Autobiography of Goethe. Melville's copy of *Goethe's Autobiography And Travels,* vol. II in Bohn's Standard Library, bears the inscription, *"Herman Melville. Bought at Bohn's, Dec. 25, 1849."*

Since he was in Portsmouth on the 25th, he evidently meant to write 24th, his last day in London. This copy, with vol. I, which bears no inscription nor markings, is in the possession of Dr. Henry A. Murray, who has generously let me use it. In his own copy, Melville marked the following passages:

I clearly felt that a creation of importance could be produced only when its author isolated himself. . . . — p. 38.
Generally, there is nothing to be compared with the new life which the

[173]

sight of a new country affords to a thoughtful person. Although I am still the same being, I think I am changed to the very marrow. . . . — p. 371.

When one enters once into the world, and gives way to it, it is necessary to be very cautious, lest one should be carried away, not to say driven mad by it. I am utterly incapable of adding another syllable. . . . — p. 450.

Besides the marked passages, there are others of great interest to anyone wishing to trace the relation of the books bought on this trip to Melville's later writing — especially to *Moby Dick.*

In the following passage he checked the sentence beginning "What are you thinking about"; I give the rest of the passage — remarks recorded of a young Captain of the Papal guard.

As I often sat quiet and lost in thought he once exclaimed *"Che pensa? non deve mai pensar l'uomo, pensando s'invecchia;"* which being interpreted is as much as to say, "What are you thinking about; a man ought never to think; thinking makes one old." And now for another apophthegm of his: *"Non deve fermarsi l'uomo in una sola cosa, perche allora divien matto; bisogna aver mille cose, una confusione nella testa;"* in plain English, "A man ought not to rivet his thoughts exclusively on any one thing, otherwise he is sure to go mad; he ought to have in his head a thousand things, a regular medley." — vol. II, p. 339.

Medal. In a letter to Evert Duyckinck, New York Saturday Evening, February 2, 1850, Melville writes —

. . . No. 3 is a bronze medal which I mean for your brother George, if he will gratify me by accepting such a trifling token of my sense of his kindness in giving me an "outfit" of guide books. It comes from a mountainous defile of a narrow street in the Latin Quarter of Paris, where I disinterred it from an old antiquary's cellar, which I doubt not connected, somehow, with the Catacombs & the palace of Thermes. — NYPL–D

It is not known which of the two medals listed this was.

"Curios" — Porcelain stoppers.

No. 6 (which brings up the rear of this valuable collection) is a bottle-stopper from Cologne, for yourself. Do not despise it — there is a sermon in it. Shut yourself up in a closet, insert the stopper into a bottle of Sour Claret, & then study that face. — NYPL–D

This letter is given in full in Willard Thorp's *Herman Melville.*

Hadjypoor. Or Hajeepoor, a village of Punjab on the Chenaub, 42 miles west of Mooltan.

London. The following list of London pleasures forms an interesting postscript to Melville's *Journal.* He compiled the list three years after his own visit for his uncle, Peter Gansevoort, who was planning a trip to Europe. He does not forget the Gansevoort love of good food and drink, nor does he neglect human interest in general.

London

Covent-Garden — before breakfast.
A sail in the Penny steamers on the river.
A Lounge on the bridges.
Go to Greenwich Hospital in the *morning* so as to see the pensioners at dinner. (An American negro is among them)
Greenwich Park, for the view.
Leicester Square — the French emigrants. Good breakfast there.
Judge & Jury — Drury Lane
Seven Dials — Gin shops.
The Temple — Go to the Church on Sunday.
    (Templars' tombs) — Dining Halls & Desert Room & Kitchen on week days.
Lincoln's Inn Kitchen.
Reform Club.
"Blue Posts" Cork Street near the Arcade — fine punch & dinners.
For fine ale — Edinborough Castle Strand —

I am indebted to Mr. Jay Leyda for a photostat of this list. — NYPL — GL

INDEX

INDEX

[186]

THIS BOOK, COMPOSED AND PRINTED IN
CAMBRIDGE BY THE HARVARD UNIVERSITY
PRINTING OFFICE, IS SET IN LINOTYPE
BASKERVILLE. THE ILLUSTRATIONS WERE
PRINTED BY OFFSET-LITHOGRAPHY BY THE
MURRAY PRINTING COMPANY, WAKEFIELD
MASSACHUSETTS.

WOOD ENGRAVING BY STANLEY L. RICE